D Y N A S T I E S

DYNASTIES

THE RISE AND FALL OF ANIMAL FAMILIES

Stephen Moss

Foreword by David Attenborough

BBC
BOOKS

Foreword 6

Foreword

Families have quarrels. Sometimes, they even have bust-ups and, as a consequence, split forever. Sociologists, had they been observing them at the time, might well have foreseen such ructions, long before the family itself realised what was happening. Animal sociologists – ethologists to give them their proper name – can often do exactly the same thing. Those studying animals that live in families, troops or herds, spend years observing such communities, trying to understand the rules that govern them, and they can, as a consequence, sometimes predict that animals are about to do the same sort of thing. What happens next can be not only dramatic but also very revealing of the nature of the animals themselves. But recording such events would be difficult.

Individual animals are often not as easy for us to identify as individuals of our own species. Sometimes ethologists have to attach radio-tags to the animals they are studying so that they are able to be absolutely certain of their identities. They also give them names so that they can easily describe what is happening – or is about to happen. Usually the names they choose have no similarity to the names we use for our children and friends. They do that in order to avoid being accused of one of the cardinal sins of ethology – anthropomorphism, that is to say, attributing human characteristics and emotions to an animal without adequate justification.

Some degree of anthropomorphism, of course, is justifiable and inevitable. If an elephant, on seeing you, lifts its trunk, flaps its ears and then charges, you are justified, at the very least, in saying that it is angry. That, certainly, is attributing a human emotion to an animal. What other word in our everyday vocabulary do we have to describe its feelings? But suppose you watched an elephant coming across a pile of elephant bones, picking them up with its trunk, one by one, as if caressing them. It would be tempting to say that the animal was mourning the death of a relative – tempting, but unjustified. Even if you know that the bones had belonged to a member of that elephant family, you could not be sure of what was in its mind. Calling this book, and the television series on which it is based *Dynasties* might in itself seem to be sinfully anthropomorphic. It will, after all, remind many of the famous American television series, *Dynasty*, which ran for many years about a human oil-rich family in the United States whose interpersonal relationships were so sensational and so fractious. Happily, however, the dictionary legitimises such use, for it says no more than that the word refers to 'a succession of rulers of the same line or family'. Animals have families, just as we do, and that is exactly what this book, like its parent series, is about.

ABOVE

Sir David Attenborough
with the pack of painted
wolves, one of Africa's
most sociable animals.

OVERLEAF

Helicopter pilot Gert Uys
flew all the aerials of the
painted wolves, including
some remarkable
manoeuvres filming
Sir David Attenborough's
introduction.

To choose their subjects, the producers consulted ethologists all round the world, asking whether the particular animal group they were studying was itself approaching one of the crises which inevitably overtake even the most amiable and well-established families. From the answers, they selected five, as varied as possible both in the nature of the animals themselves and the sort of dramas that were likely to overtake them. Camera teams then joined the scientists and followed the fortunes of each of those families for up to two and a half years.

It was a risky plan. It could be that in spite of the ethologists' predictions, nothing dramatic would happen, that the animals concerned, day after day, month after month, would continue doing exactly the same sort of thing, without any radical change. In such a case, even though they were filmed over such a long time, there would be scarcely enough incidents to justify an hour-long programme. It might also be that a crisis would lead not to happier times with a new generation, but a failure of the animals concerned to meet the demands of their new situation. But the producers determined before the series went into production that, once a community had been chosen, the drama would be told exactly as it happened.

You must now be the judge as to whether these varied and extraordinary histories are tragedies or triumphs.

David Attenborough

CHAPTER ONE

Lions

OPPOSITE

Charm, the matriarch
of the Marsh Lions,
leads her pride through
Musiara Marsh, in
Kenya's Masai Mara.

OVERLEAF

Lionesses such as Charm
are perfectly suited to
their savannah grassland
habitat, as her low, long
shape and sandy coat
allow her to blend in
with her surroundings.

SOFT RAIN FALLS across the plains of Kenya's Masai Mara, soaking trees, grass and every living thing in its path. Little groups of vultures hunch beneath the broad canopies of acacia trees, seeking shelter from the steady downpour. Giraffes and zebras, impalas and gazelles, buffaloes and wildebeest, stand around in loose herds, always alert to any danger. And hidden in the long grass beneath an acacia tree, out of sight of all but the most acute observer, a lion waits patiently for the rain to stop. Her name is Charm, and she is a mature, experienced female, the leader of her pride.

A typical lion pride usually contains between 12 and 15 lions: several adult females (normally sisters or cousins), along with between two and four adult males and their offspring, which range in age from newly born cubs to adolescents between one and three years old. However, some prides may be far bigger, with up to 30 (and in one instance, in South Africa's Kruger National Park, almost 50) animals in all. Territory size varies too: where food is plentiful, as here in the Masai Mara, it may be as small as 50 square km (20 square miles); but in dry areas it can be as large as 2,000 square km (roughly 800 square miles).

All other big cats are essentially solitary animals, the lion being the only one that habitually lives in a larger social group, whose members help one another both by hunting together and by safeguarding the cubs. So the pride is subject to complex rules, to ensure that it remains stable and sustainable.

Typically, the most mature female – like Charm – leads a lion pride. The younger females, who between them keep the pride fed and look after the cubs, then support her. The males, in turn, keep the females and youngsters safe. So although at first sight the bigger, stronger and more impressive-looking males appear to be in charge, the reality is rather different.

But Charm's pride was far from typical in that it had no fully mature males. When the team began filming, there were just ten lions in the Marsh Pride, all of which were related to Charm, and so formed part of her dynasty. As well as her cousin Sienna, Charm had four offspring: three-year-old male Tatu, the lion equivalent of a teenage boy, and her eldest daughter Yaya, likewise three years old; and two younger ones – a male, Alan, and a female, Alanis. Sienna also had four: her three-year-old son Red, Tatu's constant companion; and one younger son and two daughters from a later litter. But none of the males was yet old or experienced enough to do the duties expected of mature males.

Camerawoman Sophie Darlington sums up the role males usually play in the life of the pride – and, more importantly, what they *don't* do. "Male lions only come in handy when the pride is under threat, maybe from other lions or a pack of hyenas. They are hopeless at hunting – they're like giant haystacks, so the prey animals see them easily and run away. They're also very lazy – or perhaps I should say 'good at conserving energy' – and they are incredibly greedy, so whenever the lionesses do make a kill the males take the majority of the food."

Lion cubs are highly
dependent on the females
of the pride to look after
them and keep them fed,
warm and safe.

But there are advantages to social living for the females, too. Often, one of the lionesses will stay behind and look after the smaller cubs while the others go to hunt. Then, if they make a kill that is too big to bring back, they will call the remaining female and the cubs in to feed – that way every member of the pride benefits.

Young males are usually encouraged to leave when they reach the age of two or three, as otherwise their ability to breed would begin to threaten the older males. Once forced out, a young male will, if he is lucky, join up with one or more of his relatives, leading a nomadic existence, until eventually they are ready to take over a pride elsewhere. If he cannot find a companion, he must lead a solitary, paw-to-mouth existence, always wondering where the next meal is coming from.

Likewise, some females are excluded from the pride, either by force or of their own choice. They also become nomadic; sometimes on their own, more often in pairs (again, often joining forces with a sister or cousin in order to improve their chances of survival).

The wandering life is far from easy: if female nomads approach a pride they are usually driven away, and sometimes killed. They will then be forced to live some distance away from other lions. This often means that they have to avoid the best habitats – the more sheltered, wooded areas where they can get plenty of shade during the heat of the day – and instead live out in the open on the unshaded plains.

As programme producer Simon Blakeney points out, another crucial difference between male and female lions is that the females have grown up within the pride, but the males have come in from elsewhere. "I was fascinated by the idea that, from birth, 'success' for male and female lion cubs is so different.

Tatu greets his fellow members of the Marsh Pride. Lions are the only sociable member of the cat family, and the deep, strong bonds between the pride members require constant reinforcement.

For males, the successful ones will leave the pride where they were born; whereas for females, success is staying put – inheriting the pride from their mothers."

The leadership of a lion pride passes down the female line, from mother to daughter, for generation after generation. A pride of lions is essentially a matriarchal society – and Charm was the embodiment of a lion matriarch. But she was one with a difference, in that she did not have the support of any mature males. At times her pride was in mortal danger; facing threats from other lions, rival predators, and some of the human residents of this corner of the African savannah. Charm was the latest of her dynasty to rule, stretching back countless generations – but would she be the last?

The Marsh Lions

Charm's pride are known as the Marsh Lions, because for many years they have made their home in the boggy ground near the Mara River. The Marsh Lions have been television stars for more than 20 years, ever since they were featured in the BBC series *Big Cat Diary*.

When the episode on lions was first mooted, there was never any doubt that the team would follow Charm's pride. As Sophie Darlington, who has spent more time watching and filming lions than most, points out: "We know the Marsh Lions – we care about them – and we know they are used to people and vehicles, so don't change their behaviour when they are being filmed."

Sophie is unstinting in her admiration for Charm. "She really is an incredible lion. When you looked at her face, you could see lines around her eyes like the rings in an old olive tree; she was jowly and saggy, and built like a shot-putter. You could tell she wasn't young – her back was bowed – and yet she was so strong."

For Sophie, and the rest of the team, it was this combination of character and unpredictability that made these lions' story so compelling. They had to accept that they had no idea how things would turn out in the end.

What transpired is the story of how this one lioness battled to bring her pride back from the brink, having almost lost everything. How, through skill, experience, determination – and a certain amount of good fortune – she finally won through.

Ultimately, it is a story played out countless times in the natural world, as the leader of a group of animals fights fearlessly to secure the future of its dynasty.

Charm's story

This story began shortly after the adult males who had helped protect the pride, and fathered Charm's cubs, headed away from the marsh to conquer new territories, meet new females and produce more young. In some ways they were the victims of their own success: they had sired so many cubs that many of the breeding lionesses were actually their own daughters, which reduced the number of females they could mate with. Although they returned to the area from time to time, they played no further part in the life of the pride.

Their departure left Charm and her family uniquely vulnerable. She had helped to rule this pride for years, but never without the help and support of big adult males, who had always defended their females and offspring with courage and determination.

And without the help of the males, the pride was more at risk than ever. All around its home range, there were other groups of lions, all with at least two aggressive, territorial males. She and her fellow females – including Sienna – now had to be prepared to fight off intruders, hostile prides that would try to take over their territory. Powerful and experienced Charm may be, but she was facing the biggest challenge of her long – and until then very successful – life. How could she survive without the protection of the males?

As pride leader, Charm
had to be constantly
watchful, looking out
over the savannah to
check for any danger.

There were two likely scenarios: either the males who had recently left
would return, or a new set of males would arrive. What no one expected was
that the usually unstable situation of a pride without any mature males would
continue for so long, as Sophie explains: "We never dreamed that we would film
the pride for almost 18 months without any adult males being involved – until
right at the end."

The first thing that struck both Sophie and her colleague, cameraman John
Aitchison, as unusual when they arrived on location was the sound or, in this
case, the lack of it. "Normally, when you follow a pride, at some point during the
day or night you hear lions roaring. But with Charm's pride, we rarely heard them
at all – they hardly ever roared."

It's often assumed that only the big males roar, in order to warn other lions
to keep away. But as John points out, once the dominant male or males begin,
the other members of the pride join in – not just the males but females too, and
even the cubs. "Even the littlest cubs sometimes take part – on occasion I've
seen the tiniest male cubs miaowing as the others roar!"

But, for virtually the whole time, Charm and her pride kept almost silent.
This was because the very last thing she wanted to do was to attract attention;
with no males in attendance, the likelihood was that if they attracted another
male from outside the pride he would kill the cubs.

Lions and their lives

ABOVE

Unlike the other big cats, male lions are very different in appearance from their mates: much larger and with that huge and distinctive mane.

Being a social animal has both advantages and drawbacks: a lion must share any food it catches; on the other hand, it is more likely to be successful because lions hunt cooperatively.

Male lions are noticeably larger and heavier than females: a typical mature male weighs in at between 150 and 225 kg (330–495 lbs), while an adult female is roughly 20 per cent smaller, at 122–192 kg (268–422 lbs). The largest male lions ever recorded tipped the scales at over 270 kg (595 lbs) – between three and four times the weight of an adult human being.

Lengthwise, including the tail, a fully grown male lion can be up to 3.5 metres (12 feet) long, while a big female can reach a length of almost 3 metres (10 feet). The lion rivals the tiger as the largest of the big cats, and is the biggest carnivore in Africa.

The most obvious difference between the sexes is the huge mane sported by adult male lions, which they use during confrontations with rivals. This is made up of thick, dense fur, which frames the face and covers much of the head, neck, upper back and chest. The colour and extent of a lion's mane can vary considerably – occasionally even being wholly absent – and in general the larger and darker the mane, the healthier the animal.

Males with the darkest manes are mostly more successful in terms of reproduction and the survival of their cubs, but the drawback is that during

Male and female lions often engage in brief but sometimes quite intense scuffles with one another after mating.

hot weather the darker mane is something of a disadvantage, as it retains heat. This may, however, make the male more attractive to females, using the 'handicap principle', in which the male shows any potential mates that he is able to sport a large, dark mane and yet still be fit and healthy.

The presence or absence of a mane in a male lion is a good indicator of its age: young males start to develop their mane at about two years old, just before they become sexually mature. However, they are not able to father cubs until they reach full maturity at five years old, by which time the mane is fully grown.

Typically, given the dangers they face from fellow males and human hunters, male lions live for between 10 and 14 years, while females can live longer – up to 20 years, though they usually stop breeding around the age of 15.

Finding food

As leader, Charm was always in the forefront when the pride was under threat from attack – including from one of the biggest dangers they faced, a resident herd of buffaloes.

As apex predators, lions are at the very top of the food chain. Virtually all their prey consists of mammals – usually large grazing animals such as wildebeest, buffalo, zebra and antelopes.

The typical size of their victims ranges between 200 and 550 kg (440 to 1,210 lbs), while larger creatures such as elephants, hippos and rhinos are generally avoided. So are giraffes – their long legs can easily smash a lion's skull with a single blow. However, when the chance presents itself, lions will sometimes hunt giraffes or even young elephants.

Simon Blakeney recalls that after the wildebeest had moved on, especially during the wetter months of the year, the Marsh Pride also hunted a lot of warthogs: "They were the staple that kept them alive through the lean times – if they couldn't catch them in the open, they sometimes even tried to dig them out of their burrows. This was mostly unsuccessful though – often they just ended up getting a face full of dirt flicked at them by the furious warthog below."

A lion pride's mature females, who often work together as a team, carry out most of the hunting. Unlike other carnivores such as hyenas, hunting dogs and cheetahs, which follow their prey and then chase it down, lions are mostly 'ambush predators', waiting for their prey to come to them. They will try to identify a likely target – often a young, sick or lame animal – and then manoeuvre themselves into position to attack.

Lions can run at speeds of up to almost 60 kph (37.2 mph) over a maximum of 100 metres, but most chases are carried out over a far shorter distance – usually less than 50 metres – they let the prey get as close as possible to them before they pounce. That's because if a lion has to run any further than this at full speed it will usually overheat and lose its chance, as lions are built for power over short distances, not stamina. Even so, fewer than one in three of all hunts are successful.

It is often assumed that because lions *can* hunt communally, they always do so. But recent studies have revealed that roughly half of all hunts involve just one lioness, a fifth involve two, and the remainder (roughly one-third) feature more than two – usually between three and eight but on occasions as many as 14. Males do occasionally take part, but are often more of a hindrance than any help, as their large size and prominent mane make them easily visible to their intended victim. As a result, male lions – whether nomads or part of a pride – virtually never make their own kills; instead they steal food obtained by their fellow lions or other predators.

Lions are ambush
predators, relying on an
ability to get very close
to their intended prey by
slow and patient stalking,
before using a burst of
speed to chase and catch
their victim.

Usually the matriarch of the pride takes the lead, and unless the intended prey is particularly large she will attack on her own at first, rushing in and leaping onto the target to bring it down to the ground. Victims are strangled, the lioness grabbing her prey by the windpipe and cutting off its air supply. Her support team swiftly follows her, as they seek to neutralise any threat and ensure a quick and clean kill.

Once the animal is brought down, a bite to the throat usually results in a rapid death, though smaller animals such as warthogs and gazelles are sometimes disembowelled and eaten while still (just about) alive.

Having killed their victim, lions open up the belly and feed on the soft tissue inside, then turn their attention to the rest of the carcass. Once they have feasted on the flesh, they leave the bones and skin to the waiting vultures, jackals and hyenas.

Following a successful kill, all the pride members will usually come to feed, the dominant males literally taking the lion's share – a big male can consume as much as 30–40 kg (66–88 lbs) of meat at a single sitting. Mature females take their turn next, while old and sick animals, and the cubs, feed last.

Females need much less food than males: typically, about 5 kg (11 lbs) of meat a day although, like the males, they will eat far more if it is available, as they can never be certain where the next meal is coming from.

A Marsh Pride lioness bravely brings down an adult male buffalo, which will provide a big meal for her fellow pride members.

If they succeed in killing a large animal such as a buffalo or wildebeest, lions will not normally need to feed again for several days; which is why our typical impression of a lion pride is that activity is at a minimum.

Lions are not the only predator on the African savannah, and must frequently compete with other large carnivores such as hyenas, painted wolves, leopards and cheetahs. Under most circumstances the lions win out, either driving away or occasionally killing their rivals. However, on occasion, a lone lion may be forced to retreat, or even occasionally be killed, by a large group of hyenas.

Lions also keep a close eye for circling flocks of vultures, usually the first indication that there is a dead or dying animal somewhere close by. Like most big cats, lions are also opportunists, and will turn scavenger when the chance arises. Indeed, studies have shown that at times lions will scavenge more than half their prey, often by forcing hyenas off a kill. Yet Charm and her pride rarely attempted this, probably because they lacked the firepower of large, mature males.

But by August and September, the pride had no need to resort to scavenging, for there were plenty of animals to hunt. This time of year sees the peak of the annual wildebeest migration, as vast herds of these mighty grazing animals seek fresh grass on which to feed. In doing so they travel huge distances across the Masai Mara and Serengeti in neighbouring Tanzania. And that meant that they were passing right through the home territory of Charm's pride.

Charm and the rest of the pride feed hungrily on the remains of a buffalo she has just managed to bring down and kill.

Wildebeest travel in groups numbering tens of thousands of animals, which affords safety in numbers, and ensures that for any individual animal, the chances of being caught and killed are very small. But from time to time a wildebeest may become separated from the herd: possibly because it is sick or injured, or a newly born youngster that struggles to keep up with the others.

One day, Charm spotted an isolated wildebeest, alone on the wide savannah grassland. But that was where her good fortune ended; for this was no weakling or youngster, but a big, healthy adult bull, weighing almost 300 kg (660 lbs) – more than twice Charm's own weight.

So the odds were already stacked against her even before she began. If she tried to attack the wildebeest on her own, she might get hurt. And as with any prey animal, the chances of her spooking the wildebeest by making the wrong move were high. Even for an experienced lion like Charm, the majority of hunts end in failure.

Charm patiently stalked her prey, moving forward step by step and using the long grass as cover, until she was close enough to make a surprise attack. She grabbed the wildebeest by the loose skin of its throat with a vice-like grip, using her hind legs to knock it off balance as it struggled to escape.

It worked. Her long years of experience paid off, and despite the bull's huge strength and weight advantage, she tightened her grip until the animal fell to the ground, unable to breathe. A swift bite to the throat brought its life to an abrupt end.

Problems for the pride

ABOVE

Despite being almost fully grown, Tatu still wanted love and affection from his mother Charm – in the form of grooming his fur.

The Marsh Lions mostly live, as their name suggests, on or near Musiara Marsh, a fertile area of grass, sedges and reeds close to the Mara River. During the dry season, when areas of standing water are at a premium, the marsh attracts large numbers of grazing animals that come here to drink.

Nevertheless, Charm relied heavily on Sienna – but that presented a problem. Sienna had never been as reliable as her cousin: she was highly unpredictable, wandering away from the pride at crucial times.

And as John Aitchison recalls, another major problem arose. "Soon after we started filming, Sienna was very badly injured, and so really wasn't able to hunt; in fact, with this huge, gaping wound in her side she spent much of her time isolated from the rest of the pride."

With Sienna hurt, the whole pride was dependent on Charm to get their food – she pretty much had to hunt on her own. And as John recalls, "because she was so competent she was able to do so, and feed the pride."

Having watched and filmed so many lion prides over the years, John believes that the lionesses are not all equally suited to catching prey: "I think that some lionesses have a real ability to hunt – a greater inclination – as often

LIONS

Sienna goes missing

BELOW

At one stage Sienna went missing from the rest of her pride, and possibly in an encounter with members of a rival lion pride, she suffered major and potentially life-threatening injuries.

Three months after the adult males had decided to leave the Marsh Pride, Sienna disappeared. On the other side of this vast, flat plain, she was in big trouble. We can't know for sure what happened, but it seems likely that she stumbled across a group of hostile lions from a neighbouring territory. With no males to support her, she had to fight them off on her own, in a long and ferocious battle. Territorial fights among lions can be vicious, and Sienna was lucky to be alive – though she was seriously injured, with deep wounds on both her flanks, leaving her at grave risk of infection.

Following the fight, she wasn't able to feed for days and was rapidly getting weaker. This was a major problem: if she tried to hunt by herself, she risked making her wounds worse, and as a lone lion her chances of getting enough food to eat were slim. So unless she could recover before she starved, Sienna was doomed to suffer a slow and lingering death.

ABOVE

The huge gash in her flanks made it very difficult for Sienna to hunt, as she struggled to reach the speeds she needed to bring down her prey.

Producer Simon Blakeney was there when one of the most dramatic events in the whole story happened, as Sienna attempted single-handed to bring down a wildebeest – part of a herd that had unwittingly wandered into the area where she was resting under a tree, clearly in a great deal of pain. "As the herd drifted ever closer, Sienna realised this was a chance to get food. She waited and waited, and then seemingly from nowhere found the strength to accelerate to full speed."

The first wildebeest spotted her, turned tail and ran, spooking most of the others, but then almost immediately she spotted another, and leaped down a bank after it. "She was clinging on to its back, trying as hard as she could to pull it down, but she didn't quite have the strength to hold on as it struggled."

The wildebeest eventually managed to throw her off, and then attacked her, hitting her at least once as she tried to protect herself. As Simon recalls, this was for him the key moment at which he realised that Sienna's future was in the balance. "When Sienna didn't manage to make the kill it was clear that she wasn't going to survive if she didn't get some food soon, and she had just missed – by the narrowest of margins – the best opportunity she was going to get."

Yet Sienna would eventually overcome the odds of surviving such serious injuries.

MEANWHILE, BACK WITH the pride, Charm had her work cut out to keep her family together. Without Sienna, she had to not only hunt on her own, but also protect the younger lions from danger. The good news was that Charm's eldest daughter, Yaya, was growing up fast. At three years old, she was maturing into a lithe, powerful animal, ready to join her mother in chasing down prey.

John witnessed her gaining the skills needed by any lioness if they are to survive into adulthood: "Yaya had a real urge to hunt – she was really focused, so even when the rest of the pride were lying asleep under a tree, in the heat of the day, she would go off to try to catch a warthog on her own. She became really good, quite quickly. As a result, she soon became a really useful partner to Charm – and so at dusk they would go off together. And they had a really strong mother-and-daughter bond."

Simon Blakeney also recalls the extraordinarily close relationship between Charm and Yaya: "Even though Yaya was huge – far bigger than her mother – whenever she seemed to be confused or concerned, or in any danger, she would make a beeline for Charm. It really showed how strong the bonds between mother and daughter lions can be."

Sophie remembers that Yaya was also very curious about the world around her. "She would poke her nose into absolutely everything, and was so inquisitive – nothing was safe. Including, by the way, us!"

Wildlife cameraman John Aitchison worked closely with field guide Tash Breed to get intimate views of the pride without disturbing their natural behaviour (left). Likewise, Sophie Darlington and field guide Sammy Munene, used a special low-angle camera rig to film Charm as she moved one of her new-born cubs (right).

OPPOSITE

John Aitchison and Tash Breed filming the Marsh Pride as a storm looms overhead.

To film the lions at their own level – right down in the grass – Sophie used a special platform built on the side of her vehicle. This led to a couple of close encounters with this lioness. "One day I was filming the pride as they walked past, a little distance away, when my guide and driver, Sammy Munene, suddenly started driving off. When I asked why, he pointed down into the grass and, sure enough, there was Yaya. She wasn't exactly stalking me but, like all cats, she was very curious – she was like a big kitten!"

SAMMY DIDN'T JUST get Sophie out of trouble a few times, he also gave her an incredible insight into the lives of the lions he has been watching for so many years. "Sammy sat with me all day, and we spent hundreds of days in the field, sixteen hours a day. He was so calm and patient, and gave me so many quiet words of wisdom about lions and their lives. We simply couldn't have made this film without him."

Sammy, along with brother-and-sister team Dave and Tash Breed, had spent decades following the lions of the Masai Mara, and the Marsh Pride in particular. Not only did they spot and identify the lions more often than anyone else on the team (often relying on Tash's careful drawings of their whisker spots, whose patterns can be nearly as distinctive as our fingerprints), but they drove the filming cars too.

As John recalls, "This was off-road driving at its wildest. It was exhilarating and sometimes dangerous, for instance when heavy rain filled the gullies we had to cross, raising the water above our headlights, or during the tense hush while a bull elephant ran his trunk along the open side of our car and Tash and I held our breath until he'd stepped silently away."

A night hunt

ABOVE

Charm heads out at night, to hunt under the cover of darkness.

OPPOSITE

Lions often hunt at night: the temperatures are cooler, and it is much more difficult for their prey to see them approaching.

Groups of lions are often seen loafing about during the daytime, either asleep or simply resting, which they do for up to 20 hours a day. As a result of this apparent lack of activity, lions are sometimes considered lazy animals.

The reality is that their large size and heavy muscle mass means it takes real effort to move, especially in the fierce heat of the midday sun. Come nightfall, things are very different. As dusk falls, prides will typically start to become more active, playing and grooming before setting off on a nocturnal hunting expedition. This allows them to take advantage of the cover of darkness and the cooler air to stalk their prey.

Night-time is also when male lions usually roar: they do so between dusk and dawn to advertise their presence, warn nearby rivals, and stay in touch with the rest of the pride. A big lion's roar can be heard as far as five kilometres (over three miles) away, and is one of the classic sounds of the African night.

Like many cats, lions have exceptionally good night vision, up to six times as sensitive to light as our own. They have bigger eyes relative to their size than many other animals, and a large, round pupil (as opposed to the vertical slit of domestic cats) which allows more light to reach the back of the eye. And like

'After dark, this hunt was finely balanced between success and failure.'

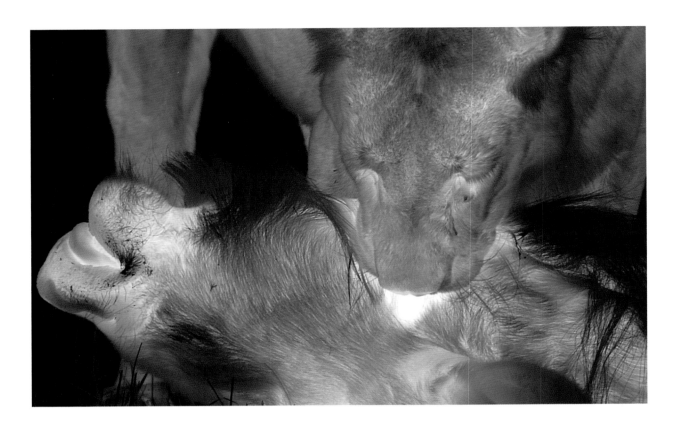

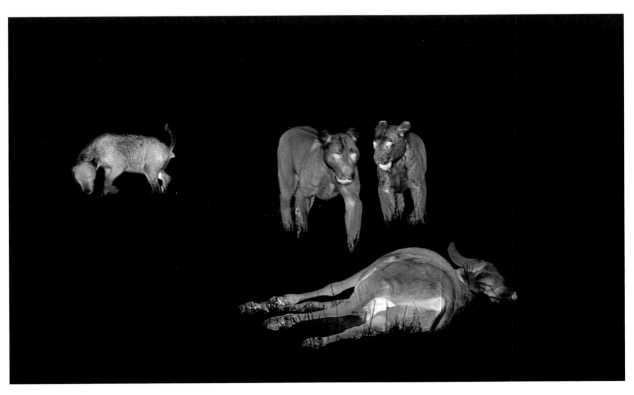

other creatures that are active at night, their eyes contain a higher proportion of rods (which are more sensitive in low light conditions) than cones, which are better suited to seeing in daylight.

While lions have the advantage of good vision, wildebeest have an exceptional sense of smell, honed over millions of years of evolution to enable them to detect and evade predators. They also have acute hearing. So, hunting after dark, Charm had to approach silently, from downwind, to prevent the wildebeest hearing her or picking up her distinctive scent. Yaya followed close behind, ready to join in if required. Charm went closer and closer until she was near enough to run in, leap up and grab an unsuspecting wildebeest. It never even heard her coming.

Things looked good: the wildebeest was gradually getting weaker, and its cries of distress were in vain: none of its fellows would ever dare to try to help it.

Unfortunately for Charm, however, she was not alone. A group of hyenas had heard the calls of the dying wildebeest, and decided to muscle in on the action, trying to grab the prize for themselves. Charm could easily be overpowered by a determined pack of hyenas.

Lions and hyenas are mortal enemies, and will always try to drive the other off, or even kill one another if they get the chance. The way hyenas attack lions is by working together. At first, they are tentative, but gradually they become more and more daring, each one dashing in when the lion's back is turned and nipping the bigger animal on their back and haunches. The lion, of course, reacts by turning to face its adversary, who swiftly retreats, while another hyena nips in behind to bite the lion once again.

Sometimes, unless a big male is present, the hyenas can eventually drive the lions away by sheer force of numbers, and they can also kill a lion if it is cornered on its own. Charm had no male to help her, so her chances did not look good – at least not at first. She could either retreat or fight – but this presented a tricky dilemma. Leaving is always the sensible course of action but, if she did so, she would lose a precious kill that would help sustain the pride for the next few days.

But soon help arrived. Yaya had followed her mother on the hunt, and came rushing in to help. Even with two lionesses, however, the odds were still stacked against them, but Charm and Yaya fought to keep their precious kill. This was a risky strategy: if one of them tripped or fell, the hyenas could easily overwhelm them. Yet it worked – the combined might of these two powerful lionesses more than kept the hyenas at bay. By the time the rest of the pride – including the bulk and muscle of Red and Tatu – turned up to support them, Charm and Yaya were already eating.

Red gets into trouble

ABOVE

On one occasion, Sienna's son Red blundered into the middle of a hyenas' den, before realising that he was in big trouble.

Like all adolescent male lions, Red did get into a spot of bother from time to time. At this age, as he approached maturity, he often wandered away from the pride, a trait that would one day become useful if he were to take over a pride of his own, and have to patrol the boundaries of its territory. But not long after the night-time encounter with hyenas, just before dawn, Red found himself alone, and this time he stumbled across a hyena den.

He had misjudged the strength of this clan. And his error might have proved fatal. Soon an angry group of hyenas surrounded him – maybe 40 animals in all.

John Aitchison witnessed the dramatic events as they unfolded: "You could see the moment when Red realised he had made a really big mistake – he was used to being with other lions, and suddenly he was on his own, surrounded by a snarling pack of hyenas. He lunged at them, but then another would bite him on his back; when he span round to see that one off, another would dive in and bite him again – so he was getting weaker."

As John recalls, "It was a desperate standoff as he twisted and turned from side to side, lunging at each approaching hyena in turn, and snarling at his

Lions may be bigger and stronger than hyenas, but a determined pack can still cause trouble for a lone lion.

attackers. But every time he did so, yet another hyena rushed in to attack him from behind. Sensing victory, they were getting braver and braver. He would only be able to keep them at bay for so long, as soon he would become too exhausted to fend them off. Of course he should have just turned tail and ran away, but he didn't have the experience to realise that."

Fortunately, his ally – Charm's son Tatu – had heard Red's desperate calls for help. He would never abandon his cousin as their bond was so strong. From here on, the odds rapidly switched in the lions' favour. Even for a pack of hyenas, a pair of male lions is simply too much to take on. The attackers soon backed down, and sloped off to lick their wounds.

Tatu's arrival came just in time – and Red was saved. After the confrontation was over, the two cousins walked along side-by-side, rubbing their manes together in a special moment of intimacy. This is how male lions form deep bonds which, when they have their own pride, will be crucial in helping them to survive. Thanks to Tatu, Red had a very lucky escape, as Sophie remembers. "Had Tatu not come back he really could have been killed. For the rest of the day he just sat and licked his wounds – he was in tatters."

Sienna recovers and returns

By early October, it was six weeks since Sienna had been injured and become isolated from the rest of the pride. Back on the other side of the plain, her plight was getting more and more desperate.

She had mostly given up trying to hunt, resorting instead to scavenging old kills and other dead animals, which she then often had to share with the hyenas. This was keeping her from starvation, though she was still very thin.

Slowly, however, Sienna began to recover. Though her wounds had been incredibly severe, by now they had almost healed, and she was managing to scavenge more and more food each day. Little by little, she was becoming stronger and able to walk further. Finally she was getting strong enough to be able to re-join the pride.

At first she approached slowly and tentatively, perhaps afraid of being rejected, as does sometimes happen. She had been isolated from the rest of her relatives for weeks now, and they would take time to rebuild their partnership. But after a wary and tentative greeting, Sienna was welcomed back into the fold. The bond between Charm and Sienna was far too strong to be broken by a few weeks' absence.

Lions are always much stronger together than apart. They often hunt as a team, which gives them a secret weapon against their victims. The bonds within a pride – between the lionesses, and between the lionesses and their cubs – are incredibly strong, and help secure the future of their dynasty.

Charm had kept her family alive through tricky times, but now, with Sienna's return, her pride was back together. Well, almost... for not everyone was there for the reunion. Charm's youngest son, Alan, had always been the greediest member of his family, and the last to leave at the end of every meal.

Like his older brother and cousin, Alan was always getting into scrapes. One day, after eating his fill, he had fallen asleep with a full belly. But when he awoke, the rest of the pride had gone. As with Sienna, a lion pride could not afford to wait – so now Alan had to catch up the rest of his family on his own. Fortunately, like all lions, he had an amazingly acute sense of smell, enabling him to follow their progress across the savannah.

Eventually, he did succeed in tracking them down, and the Marsh Pride was complete again. And not a moment too soon, because the rains were by then long over, and the land was beginning to turn very dry and dusty, making it harder for the pride to find food. The next few months would be absolutely crucial for the survival of Charm's dynasty.

Threats to lions

Today, lions are mainly found in Eastern and Southern Africa, apart from small and isolated populations in West Africa and Ethiopia, and a tiny remnant population of Asiatic lions in the Gir Forest of northwest India, numbering just 500 individuals. Estimates of the total population of lions in Africa are hard to come by: the usual figure quoted is between 16,500 and 30,000 animals, with the best guess being fewer than 20,000 – compared with as many as 400,000 in 1950. And, in reality, there may be far fewer lions than we think.

Yet just 10,000 years ago, about the time that human hunter-gatherers were beginning to settle on the land and become farmers, the lion was one of the commonest and most widespread large mammals in the world. Lions ranged throughout the whole of Africa (a separate race, the Barbary lion, was found across North Africa including the Sahara Desert), and much of southern Europe and Asia, with an outlying population in the Americas. Before then, around 12,000 to 14,000 years ago, they even roamed across what is now Britain.

But once human beings settled down from their original hunter-gatherer existence, and began to keep and farm livestock, lions became not just a dangerous predator but also a rival, and their days of dominance were numbered. Gradually their range began to shrink, and they had disappeared from Europe by the first century AD.

Numbers in Africa continued to remain high until the start of the twentieth century, when increased pressure from hunting, combined with habitat loss and persecution by farmers and landowners, pushed down the population to its current precarious status. As a result, inbreeding and loss of genetic diversity has become a major problem. This is especially true for isolated groups such as those lions living in West Africa which are cut off from the rest of their peers, and may now number fewer than a thousand individuals.

Today lions are still widely persecuted, and with the increase in the human population and the need for more land to graze animals and plant crops, the lion's future as a species is far from assured. And it was this that almost led to the Marsh Pride's downfall.

BACK IN THE Masai Mara, the land was beginning to turn drier and drier, as the last of the rains were over. On cue, it was time for the vast herds of wildebeest, which had sustained Charm and her pride for the past few months, to head across the border and into the Serengeti, to pastures new. These sturdy beasts would not be back here for nearly a year.

So even with Sienna's timely return, and Yaya's growing strength and guile, the pride was always going to find it much harder to find food. They would need to be less choosy, and more inventive, when it came to finding new and different prey. During the dry season, lions will often switch their attention

to smaller animals such as Thompson's gazelles, monkeys and especially warthogs – anything to sustain them through this long and difficult time.

Sometimes they were forced to hunt larger prey, though this was a dangerous option as it might leave them open to injury or death. So two months later, after eking out a difficult existence by hunting smaller prey and scavenging, Charm and Sienna decided to tackle a rhino – despite the obvious dangers.

Hunting doesn't come much more difficult than this: the rhino's huge horn can easily impale a careless lion, so at first the hunters kept their distance. They changed tactics, too: instead of stalking their unwitting prey and then pouncing from close by in an ambush attack, they adopted the tactics more often used by painted wolves, and pursued their target over a longer period of time, out in the open.

But, despite their best efforts, the rhino showed absolutely no signs of weakness. Even when they finally managed to get the beast cornered, it proved surprisingly agile, twisting and turning to fend them off as they tried to attack. Eventually, having assessed the likely dangers against any potential benefits, Charm decided to call off the hunt. The cubs would stay hungry for another day, at least.

In desperation, Charm led her pride to the edge of the Mara Reserve, where there was a much easier source of prey – but one that might have long-term repercussions for the pride's survival: cattle.

CATTLE ARE SLAP-BANG in the typical range of a lion's prey – the size and shape of an antelope, but far easier to catch as, unlike wild herbivores, they have not got the benefit of millions of years of co-evolution between predator and prey.

Lions and cattle farming have co-existed in East Africa for generations. Traditionally, the cattle would be rounded up at dusk by the herdsmen and put into a compound, which meant that they could easily guard them against any passing lion. But now things have changed: some herdsmen look after hundreds of cattle, which they cannot easily guard at night.

The problem is made worse because the grass is so good inside the reserve that people risk bringing their cattle directly into the protected lion territory, often after dark. But bringing cattle all the way into the reserve when lions are hunting is a recipe for disaster – the cattle are easy prey for the lions. And sometimes the herdsmen can retaliate against the lions.

That's exactly what happened to Charm and her pride. After several nights of killing cattle, some of the lions discovered a freshly dead animal, and took advantage of the free meal – a decision that would ultimately prove fatal.

"One day we were driving along with our expert guide, Dave Breed," recalls John Aitchison, "when he saw something that struck him as very unusual. He'd

BELOW

Domestic cattle are often easier to catch than wild animals. Across Africa, this has put lions in direct conflict with herdsmen.

seen some elephants behaving very strangely, sprinting down the hill near the edge of the reserve with their ears down, as if they had seen something really frightening. Then fellow cameraman Mark MacEwen and his driver found an older lion, staggering and frothing at the mouth, and called us over to see. As soon as we arrived, Dave knew that she had been poisoned." Immediately, vets from the Kenya Wildlife Service were called.

It appears that at least one of the carcasses of the cattle Charm and the pride had fed on had been laced with poison, most likely an illegal pesticide called Carbofuran. The rest of the pride were also staggering along and looking terrible. Alan, who was always the greediest of the group, looked the worst: possibly because he had eaten so much of the poisoned meat. Even Charm looked really ill, but was managing to stay on her feet. Alan was behaving really strangely. Then he

simply lay down in the hot afternoon sun, while the others managed to reach the sanctuary of a nearby tree where they could try to recover.

When Charm looked back, it was obvious that Alan had been unable to follow them. She faced a dilemma. Her son was desperate for her help; and yet the rest of her pride needed her too, if they were to survive. She, too, had been poisoned, and could hardly walk straight.

OPPOSITE

After Alan fed from a poisoned cattle carcass, he became too weak to walk. Although Charm stayed with her son for as long as she could, eventually she had to return to the rest of the pride. Despite the vet's best efforts, Alan died soon afterwards.

At first, her maternal concern won out: by the next morning she had returned to Alan. But he was still very weak – slumped on his side in the long grass, unable to move. Charm waited and watched for a while, but it was soon clear that Alan was far too weak to move any distance. She tried walking away, then stopped and turned to encourage him to follow her. But he could not. Eventually, the need to protect the pride – including her daughters – won out over her maternal instincts to stay with Alan, so she turned and left without him. After that moment she would not see him alive again. The vet returned the following day, but there was nothing he could do to save Alan.

For the team filming the lions, who had inevitably developed a deep affection for their subjects, Alan's death came as a real blow. But John draws a wider lesson from this awful event: "If one animal represents the African wilderness it is lions; and if this is happening to lions in the most-protected place in the world, where they are watched by more people than anywhere else, then surely that's the end of lions. They are simply doomed."

"The way Alan went was absolutely brutal," says Sophie, "but it's important to know what happened. Wildlife filmmakers have been filming lions for years, and yet we haven't really ever admitted that humans are involved, and are doing the lions harm. Now we are watching the lions disappearing in front of our eyes – and it really is time we did something to stop it. If we can't stop the killing, then they'll be gone – and sooner than we think." Sienna, too, had feasted on the poisoned carcass; and she, too, was unable to recover. A few days later, the remains of her body were found, having been scavenged by hyenas and vultures. Now, with two of the most important pride members dead, Charm had lost her key ally, Sienna, and her young son, who had he survived would one day have taken her bloodline into new territories.

But there was some good news: Charm's daughters Yaya and Alanis, her eldest son Tatu, and Sienna's cubs, including Red – had all survived the poisoning. They would now be the future for Charm's dynasty. Despite everything that had happened, she had to focus solely on keeping them safe. They needed her skills and experience more than ever.

Tragically, the deliberate poisoning of lions is becoming more and more common throughout Africa, as the human population increases so rapidly, and demand for grazing land continues to grow. Across the continent, conservation groups are working to find ways that allow lions to co-exist safely with human populations – and their livestock. Any solutions must find a balance between the needs of the local people and that of the lions themselves, so that both are able to thrive, and hopefully benefit from one another.

The young males leave the pride

From then on, Charm kept the pride constantly on the move, always looking for the chance of a successful hunt. But with the wildebeest long gone, catching enough food to sustain the whole of the pride was harder and harder, even with Yaya's help. Yet somehow Charm always managed to find them something to eat; her years of experience meaning that she was just about able to keep her family fed, even during the hardest times.

Then the dynamic of the pride changed yet again. For several weeks, the two young males, Red and Tatu, had been wandering further away each day. They were now nearly four years old: almost fully grown, with increasingly dark and impressive manes. If they were ever to father cubs, and continue Charm and Sienna's dynasty, they had to leave. So one day, instead of wandering off and then eventually returning, they headed away for ever.

ABOVE

After taking over the Marsh Pride, the two new males sat together looking out over their new territory.

From now on they would live as a tight duo. They were no longer those impetuous adolescents who had tried to bring down a hippo; now they were a powerful partnership, forged under Charm's protection.

But as time goes by, the urge to father a dynasty of their own will begin to dominate their actions. Eventually they are likely to fight other males for a territory and, if they win, will form a pride of their own. Success is far from guaranteed: an incumbent male lion – or more often a pair – will fight long and hard to retain control of their pride, and so may be able to fend off any approach from outsiders – even those like Red and Tatu, who have inherited their mothers' strength and guile.

Although the loss of the two young males may appear to have been a setback for Charm, in the wider scheme of things this represented a great success. Through Tatu, her dynasty now had the chance of carrying on elsewhere, even though she was unlikely ever to see him again.

New males arrive

Early one morning, a sound echoed over the marsh that Charm and the pride hadn't heard for well over a year. It was loud, deep and penetrating, as if the whole landscape was shuddering in fear: the roars of two adult male lions.

But this wasn't the return of the prodigal sons Red and Tatu; nor was it the pride's original males returning. These were two unknown young lions that had come from way beyond where the cattle graze. For at least a year, since they left their birth pride, they had been roaming across the Masai Mara, looking for a new place to live.

Now – in what for them was a huge stroke of luck – they had stumbled across a pride without any incumbent males to challenge them. So they could simply walk right in and take up residence, claiming the pride and its territory

Solitary lions

Charm and her core pride were not the only lions in the area: as well as occasional males passing through, there were also other lionesses. Just as filming started, some of the other Marsh Pride females broke away from the core pride and formed a loose association that, over the coming months, was often found on the fringes of the range.

One of the lionesses in this peripheral group was Bibi. Well known to many TV viewers from *Big Cat Diary*, Bibi was probably the oldest surviving lioness in the Masai Mara at the time the team was filming Charm's pride there. She was also one of the easiest lions to identify. "She had the weathered look of an old lioness," said producer Simon Blakeney. "She had lost the tip of her tail many years earlier. It made her probably the most recognisable lioness on the Mara."

Despite her advanced age, Bibi mated with a passing male and later gave birth to a single cub. The odds were stacked against her from the start: at 17 she was old for a breeding lioness and, because she was no longer part of a tight-knit pride, she could not rely on other lionesses to help look after her cub when she went hunting. She also had no males to protect her and her cub against intruders or predators. Indeed, for a few days, when Bibi went hunting she actually hid her cub away in a toilet shack by the local airstrip.

She tried very hard to bring up the cub successfully, moving it whenever she was worried that it might have been discovered or be under threat. She even walked a huge distance across the baking-hot plains to take the cub to a stream bed where she had secreted many litters in the past, right in the heart of the Marsh Pride territory. Other times she took it to isolated islands in the centre of the marsh, where it would be safe from discovery.

But keeping it safe on her own was very difficult. One day the crew found Bibi, but her cub was nowhere to be seen. Bibi appeared to spend days looking for it, but the crew never saw the cub again. It seemed likely that it had been taken by hyenas, as large numbers of them had arrived in the area to feed on a nearby elephant carcass.

A few months later – ever the opportunist – Bibi was drawn by the offer of a free meal, and ate from the same poisoned carcass that caused the deaths of Alan and Sienna. Bibi, one of the oldest and most famous lions on the Mara, was also killed.

One of the two cubs born to solitary lioness Kabibi, who managed to raise them both against the odds.

KABIBI BROKE AWAY from the Marsh Pride at the same time as Bibi and formed part of the loose satellite group along with some of her cousins and litter mates. Nobody is entirely sure whose cub Kabibi was, but it seemed likely that she was one of Charm's, possibly even Yaya's sister. Kabibi was not much more than a juvenile, only about two-and-a-half years old.

About three months after they left the main pride, Kabibi and a couple of other young lionesses mated with a nomadic male, who later disappeared. Three lionesses had cubs – their first litters – and whilst the other females seemed a little bewildered by their new arrivals, Kabibi was committed from the beginning and was the only one to keep her two youngsters alive for longer than a few months.

The crew saw Kabibi with her cubs when they were only a day or two old – the youngest cubs they saw throughout the whole of the filming. Initially, they spotted her as she disappeared into a dense, forested area close to the river, looking heavily pregnant and uncomfortable. After careful observation, the crew then spotted the two cubs, whose eyes were only just beginning to open. Incredibly for a young, first-time mother, Kabibi had no fears bringing the cubs out when the crew were there.

Kabibi was a very attentive mother, only leaving the cubs for long enough to feed, and always returning to let them suckle. She was also very patient with them. "She was incredibly tolerant of the cubs' antics," says Simon Blakeney. "They would climb all over her, chewing on her ears and tail, and never staying where they had been put."

Kabibi's maternal instinct didn't only extend to her own cubs. She even fed cubs belonging to some of the other young lions, when their less-committed mothers were away. When they first tried to suckle, she would snarl at them, a

Kabibi feeding her two young cubs, while three hungry youngsters from another female try to muscle in. Despite her youth (she was just two-and-a-half years old) Kabibi proved herself to be a natural mother.

protective mother making sure her own cubs had enough to eat. But some of the cubs had been left alone for as long as two days, and they were starving. Once Kabibi's own cubs had finished suckling, she was happy for the others to take a turn. Such behaviour is unusual, but far from unknown – the maternal instinct is stronger in some lionesses than others.

Despite Kabibi's help, the other females lost their cubs, through accident or inexperience. As the only lioness with cubs now, Kabibi became isolated and, later on, the arrival of the new Marsh Pride males in the area meant that she had to be exceptionally careful that she and her cubs weren't found. If they had been, they would have been killed by the new males. As a result, she and her cubs were forced to live alone, moving around a great deal.

It wasn't easy for her. On one occasion, Simon watched Kabibi trying hard to hunt for a whole day, but all the while she was continuously followed by two barking jackals and a herd of snorting Thomson's gazelles, both of which made sure that any potential prey in the area knew exactly where she was. She had to return to her hungry cubs that night without having caught anything. But Kabibi never gave up, and she became an accomplished hunter and an even better mother. On one occasion she even managed to bring down an adult male zebra all by herself – a huge meal for her and her cubs.

Over time, the cubs grew and grew. Once they matured, they eventually joined up with a large coalition of wandering males. Along with two other females who left the main Marsh Pride when the new males arrived, Kabibi and her cubs were on the cusp of founding a brand-new pride of their own. And that was all down to the perseverance and natural maternal ability of this incredible lioness.

Charm mates with a new male

For Charm, it was finally time to begin the process of rebuilding the Marsh Pride – with the new males. This would be her best chance to begin a new line in her dynasty. So just as the wildebeest herds returned, streaming across the Mara River, she and one of the new males moved a little way away from the rest of the pride.

Sophie Darlington saw a real change in Charm's behaviour at this crucial point in her life. "She went from being old and tired to young and vivacious. Yet she remained the boss – she welcomed that male into her territory on her own terms".

Mating in lions is a long, complex and sometimes bizarre process, full of apparent aggression between the male and female. It begins when the lioness starts to come into season, which she does every eight weeks or so. During this key period she emits tell-tale scents that signal to her potential mate that she is ready to mate. From now on the male will stick to her like glue for up to six days, to ensure that no other male gets a look-in.

The couple then copulate – not just once, but as many as several hundred times in a single marathon session. The sexual act itself wins no prizes for tenderness: the male will often grab the female by the scruff of her neck and, as he withdraws, shafts on his penis roughen the inside of her vagina, which may stimulate her to ovulate but also causes pain. This can be seen from the female's typically aggressive response after each bout of mating.

But for Charm, there was a lot more to this than the pure mechanics of conception. She had another crucial aim: to reinforce her bond with the male – for the sake of her future cubs. Infanticide of cubs is very common in lions – when new males take over a pride, they usually kill the offspring of their vanquished predecessors, which ensures that all future members of the pride are related to them.

A male lion will, on occasion, even kill his own offspring – if he cannot be absolutely sure that he is the real father. So Charm knew she had to stay close to him throughout the courtship and mating period, giving him absolutely no doubt that any future offspring were his. Hopefully, when any new cubs are eventually born, he and, just as importantly, his fellow male should accept and defend them. The future of the dynasty now rested as much on his shoulders as hers.

The cubs are born

One morning Charm was lying on her side in the grass, with her pride around her, when she suddenly growled, startling the other lions that had been asleep. She was in labour. Soon afterwards she slipped away to the marsh to give birth. This was now the most critical time for the future of her dynasty.

Lion cubs, which are born about 16 weeks after conception, are much smaller than you might expect. At birth, Charm's pair of offspring – a male and a female – weighed just 1.5 kg (3.3 lbs) each, a total equivalent to about one-sixtieth of her body weight (a typical human baby is about one-twentieth of its mother's weight).

Like many mammals they were born unable to see, with their eyes not opening until they were roughly a week to ten days old. During the first month of their life, their mother moved them several times to avoid discovery: she did so by picking them up one-by-one, gently around the neck, and then transporting them to their new home.

Charm fed them on her milk, but she had to eat too, so from time to time had to leave them temporarily so that she could hunt or scavenge for food. This

BELOW

The Masai Mara at
sunset, looking towards
the Oloololo Escarpment.

---

is always a dangerous time for baby lions: any predator that finds them, from eagles and pythons to leopards and hyenas, will attack and eat them; even other lions will not hesitate to make a meal of small cubs if they stumble across them.

So even hidden away as they were, Charm's two little cubs were still incredibly vulnerable. Charm had chosen to make her den in the middle of the marsh – in retrospect a poor decision, as by then it was getting very dry, so herds of buffaloes and elephants regularly came into the boggy area to graze. If they then came across Charm and her cubs they would not hesitate to trample them to death.

This would not simply be aggression for its own sake: buffaloes are clearly well aware of the danger they face from lions, and so do their very best to destroy the new generation of lion cubs by ramming their huge horns into the long grass and attempting to crush the youngsters.

Buffaloes can and do kill adult lions as well. Yet even when she and her cubs were surrounded by these mighty beasts, Charm stood her ground, willing, if necessary, to lay down her life for her tiny offspring. There was a tense standoff. Time and again the buffaloes approached; time and again Charm defended her den and cubs against attack. Eventually, the buffalo herd retreated, and the cubs could finally come out of hiding. They gambolled playfully around their mother, all tension and danger now forgotten, for the moment at least.

It would be another two months before they could start to eat meat, but already the cubs were growing fast as they were getting all the nutrition they needed from their mother's rich milk.

By the time their eyes had opened, they had become far more active, often playing with one another: romping, stalking and play-fighting in ways that will prove remarkably useful in their future lives. These little lion cubs – one male and one female – were the opportunity Charm needed to restart her stalled family line.

A new generation

Life for lions is never easy – but these two tiny animals had a powerful, experienced mother to look after them. And now the pride also had two big males to protect it – males who both had a stake in its future; for the other new male had mated with Charm's daughter, Yaya, so that she was pregnant.

One very special day, roughly six weeks after the cubs were born, Charm brought her offspring back to meet the rest of the pride for the very first time. She approached the big males tentatively, but her fears were groundless: they readily accepted the youngsters. Having done so, the males would then defend them as fiercely as Charm – for they were the future of their dynasty too.

Charm was also reunited with her eldest daughter – and now hunting companion – Yaya. They had been apart for much of the time since Charm had given birth, but clearly the bond between mother and daughter was as strong as ever.

The cubs were, understandably, shy and nervous at first; but after a few hours they began to settle down. With luck – and under Charm's care – they now had a good chance of survival.

Charm's new cubs were the culmination of over 18 months of struggle for her and the Marsh Pride. She is a survivor – as she has proved time and time again. Through a combination of bravery, strength, guile, intelligence and caution, this extraordinary lioness has ensured there is a future for her bloodline, her pride, and her dynasty.

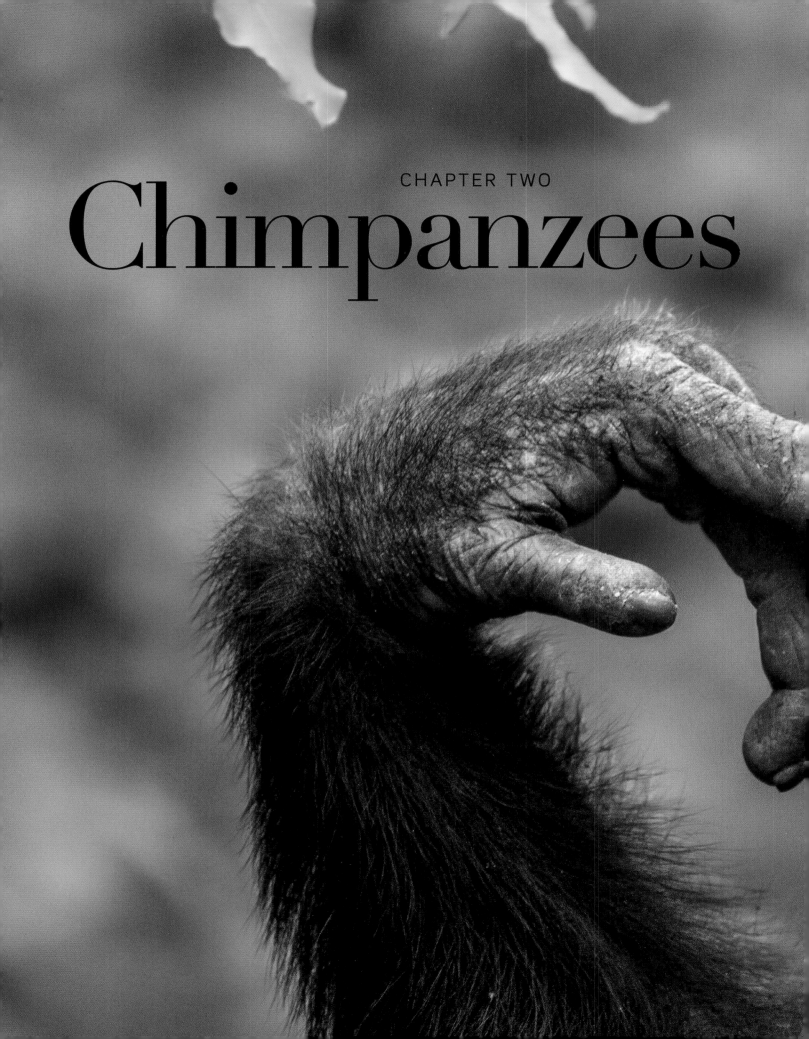

CHAPTER TWO

Chimpanzees

A highly experienced animal, David was in pole position in the group.

Luthor, a young chimpanzee, was always on the lookout for ways to rise up the hierarchy.

IT IS A baking hot day at the beginning of the dry season, in a lightly wooded patch of savannah in Senegal, West Africa. A group of chimpanzees walk slowly between the trees in search of food. They appear calm. Yet amongst this group, as in any other, there is always tension simmering beneath the surface.

The alpha male in this group is called David. He has complete dominance over all the other chimps, which gives him first access to the three crucial elements in any male chimpanzee's life: food, water and females. But uneasy lies the head that wears the crown. Like any animal in charge, David has learned to be wary. There are few, if any of his fellow males he can trust; for they want to become the alpha male, with all the privileges and benefits that brings.

Life in a group of chimps is in a constant state of flux, with the shifts in the relationships between the individual members key to what will happen next. At some point, at least one of the other males in David's group will try to bring him down. If they succeed, he could even suffer either exile or death. The stakes simply could not be higher.

At the start of the filming, David had already been in power for three years, ever since he managed, with the help of his brother Mamadou, to overthrow the previous leader. Whenever David needed to consolidate his power, Mamadou fought strongly alongside his brother to ensure that David stayed at the top.

This bond between brothers was crucial – but Mamadou then suddenly disappeared, and was never seen again. That meant that David was more vulnerable than he had ever been, and had to constantly display his power to assert control over the other males, to stop them becoming dangerous rivals. This was exhausting work, but he had no alternative if he wanted to retain his position at the top.

The alpha male in any group of chimps is always wary of potential rivals who might threaten his rule.

Most important of all, he had to stay as alpha male for long enough to ensure that his dynastic line would continue down the generations, long after he is dead. If he were able to sire a future leader of the group, that would be considered a success.

But for now, David's position was potentially under threat, from up-and-coming males in the group, particularly an experienced rival, Jumkin, and a younger one, named Luthor. Chimpanzee society is very fluid, with alliances constantly being formed and broken between different sets of males. So, although things appeared calm for now, the dynamics of the group could alter in an instant. David had to remain constantly watchful.

Even from an early age, chimps are able to use their forelimbs to climb through the trees.

different from their forest-dwelling relatives, being slightly smaller in stature, and with less body hair.

This is not an easy place to live. Most chimps live in equatorial rainforests in Central or Eastern Africa, where the climate remains more or less the same all year round. This, in turn, provides a constant and reliable supply of food in the form of many different types of fruit – forest chimps regularly feed on several hundred different kinds. But here, in this unusually open setting, David and his group have to cope with huge seasonal changes, which push them to the very limits of their endurance.

Here, the dry season can see temperatures soaring well above 40 degrees Celsius, drying up all but a handful of small waterholes, making food very difficult to find.

How chimps live

Chimpanzee society is patriarchal. Male chimpanzees are dominant over the smaller, less-aggressive females; and like many other social mammals, each community has a clear hierarchy, led in this case by the alpha male.

However, the alpha male is not necessarily always the strongest individual in the group: intelligence plays a major part in determining which male becomes dominant, and especially how long his reign lasts. Brains are at least as important as brawn, especially if an alpha male chimp is to reign for a good length of time.

As an alpha male, David regularly shows his dominance over the others by overt displays of power – using aggressive body language to induce submission in his rivals. But he also does so in less obvious ways, such as making alliances with others in the group.

But the power struggle is not entirely focused on the alpha male: females also have their own hierarchy, and in exceptional circumstances may join forces to get rid of an alpha male. The hierarchy is not just for symbolism or show: the highest-ranking females get to feed first, and get preferential access to the alpha male when they come into season and are ready to breed. Lower-ranking animals – both males and females – display submissive behaviour, such as offering their hands in supplication, or turning around and showing their rear end.

The other major difference between male and female chimps is that the males in a group have been born there, whereas the females are incomers.

Young chimps spend the
first two or three years
of their life dependent
on their mother.

Once they reach sexual maturity, sometime between 9 and 14 years old, females almost always move away, and find another group to join; whereas the majority of males stay within the community they were born. This makes the relationships between males absolutely crucial.

Each chimpanzee community has its own 'home range': an area of forest habitat covering around 7 to 32 square km (around 3 to 12 square miles), though chimps living in grassland habitats have larger ranges, up to 65 square km (25 square miles). David and his group have a really large range, extending at its extremes to as much as 90 square km (35 square miles) – because the lack of easily available food and water, especially during the dry season, means they need more territory in order to search for these.

Their foraging behaviour is very different, too, from that of forest chimps. While a typical forest group will travel just two or three kilometres a day in search of food, this group generally moves at least five km, and sometimes as far as 15 km (9 miles), in a typical day – most often in the dry season when food is scarce. Even in the times of plenty, during the wet season, chimps must feed for several hours every day.

They are often able to feed for a whole day or more in a single place, especially when a particular tree comes into fruit and produces a temporary glut. But this is rare: more often than not, they must walk long distances between one food source and the next. They do not necessarily stay together to do so; indeed, the main cohort often splits up into smaller groups, which then re-join each other at a later date, creating what scientists have described as a 'fission-fusion' existence.

The chimps' nomadic lifestyle, combined with the very high temperature during much of the year, made filming David's group even more tricky than usual. "Chimps are one of the toughest subjects in the natural world to film," wildlife cameraman Mark MacEwen recalls, "because, despite the intense heat, they would often walk many kilometres in a single day, starting at 4 a.m., well before sunrise, and finishing at nightfall, when they finally go to rest. It became a real battle of our physical and mental resources just to keep up with them."

The large size of these chimps' territory may also explain why there is an imbalance in the sex ratio between males and females here. Normally, when they reach maturity, females transfer between groups – a risky business, as they may not be accepted by other, unfamiliar chimps. The much greater distance than normal between this and any other group has probably proved too much of a barrier for lone females to cross, so they have not been regularly replenished with females from outside.

A group of chimps
being filmed.

Filming the chimps –
which often roam widely
as they search for food
– was a real challenge for
the camera team.

time guarding the borders of their territory, as there was no real need to do so on a regular basis.

However, on occasions, the group would wander as far as the banks of the Gambia River, at the very northern extreme of their range. Then, as they called to one another, the sounds they made would echo off the far bank, a few hundred metres away. This gave the impression that chimps from another group were calling back at them. As John recalls, "One individual, named Lex, would get quite agitated by this, and have a prolonged screaming argument with the echo of his own call."

DURING THE COURSE of the filming, the chimps formed a special bond with Rosie Thomas and her team. "On the first day of every trip, David would always come and walk really close by us – just letting us know that he knew that we were there. It was almost as if he was reminding us that he was the one in charge – which of course he was. After that, he would let us follow him, most of the time, for the rest of every trip. But when he didn't want us around there was absolutely no chance of keeping up with him, and we had to accept that some days he wanted to be alone."

Filming chimps, as with any wild animal, requires a healthy dose of respect for the creatures' power, strength and intellect. The crew knew they must always keep at a safe distance, never make sudden movements or loud noises, and if it becomes clear that they don't want you around, beat a retreat. Fortunately, the team had local trackers on the ground who knew this group of chimps intimately – Michel, the field assistant, could even tell individual chimps apart by listening to their calls.

Over time, the filming team realised that they could tell individuals apart from one another by particular characteristics. "For example, Mike had a slight ginger tinge to his fur, Luthor had sneaky eyes, David had a natural charisma," explains John. "These 'softer' clues are what we ended up relying on to identify individuals, much as we can recognise familiar human faces." In addition, many of the males were scarred, sometimes with fingers or parts of their ears missing, because of conflicts within the group, which also made them easier to tell apart.

Chimps and their relatives

ABOVE

Chimpanzees are – like human beings – primates, one of eight species of Great Apes.

OPPOSITE

Close relatives of the chimpanzee include the bonobo (top left), the mountain gorilla (top right) and orang-utan (bottom).

The chimpanzee is, like us, a primate: the most highly evolved group of all the world's 5,000 or so species of mammals. In all, there are roughly 400 species of primate, including lemurs and lorises; galagos and gibbons; marmosets and tamarins. But only eight are members of the family Hominidae, known as the 'great apes'.

Apart from humans, these are the western lowland and mountain gorillas, three species of orang-utans (the Bornean, Sumatran and Tapanuli – only named as a separate species in 2017), the chimpanzee's smaller cousin the bonobo and, of course, the chimpanzee itself – widely considered to be our closest living relative, having diverged from humans roughly six million years ago.

Chimpanzees display several key traits that we once considered unique to humans, including the use of tools and rudimentary forms of sign and verbal language. They also have a highly developed social structure, including the ability to hunt cooperatively.

Also like us, they are very dextrous: well able to delicately manipulate objects with their hands, fingers and thumbs – and, unlike us, with their feet and toes as well. And they have relatively big brains – with a volume of roughly 300–400 cubic centimetres. This is roughly one-third to one-quarter the size of our brains, which means they are able to solve quite complex problems.

Facial expressions are also very important in chimp society, as they are for humans. The commonest ones include a hostile expression, with the hairs raised, used during conflicts with a rival; a relaxed 'play face', in which the top

> ' Facial expressions are also very important in chimp society, as they are for humans. The commonest ones include a hostile expression, with the hairs raised. '

OPPOSITE

When chimps bare their teeth in what looks like a grin, this is actually a sign of fear.

lip conceals the upper teeth, used when two young chimps are indulging in play fights; and a series of what look like pouting faces, mostly used when one chimp is begging for food from another. The well-known 'grin', in which a chimp opens its mouth and bares its teeth, is not, as sometimes thought, a sign of pleasure or happiness, but instead indicates fear.

These particular chimps, whose habitat is so much more varied, and their lives in some ways more complex, than their forest-dwelling relatives, have attracted particular interest from zoologists and anthropologists for many years. They seem to show many of the traits of early humans, having descended from the trees to the more open plains, being able to use tools, and spending more time on two feet.

Like the other non-human great apes, chimps walk quadripedally – on all four limbs – folding in their fingers and placing the knuckles of their forelimbs on the ground. To enable them to do so without injuring themselves, their knuckles are protected by a ridge of bone, to prevent the wrists from buckling under the animal's weight. This particular group, based in an open, savannah habitat, used this mode of travel most of the time.

Other populations of chimps are more at home in the trees: using their powerful forelimbs to swing from branch to branch with incredible speed and skill. They can hang from any one of their four limbs and, like us, have opposable thumbs, enabling them to grip firmly onto any branch. Unlike us, however, they also have opposable toes on their hind limbs, a real asset when climbing. And they have one final trick up their sleeve: when necessary (for example when carrying items of food) they can, like us, walk bipedally (on two legs) – but only for relatively short distances.

APART FROM THE bonobo (a scarce cousin of the chimpanzee found only in Central Africa), the chimp is the smallest of the great apes. Males are roughly 10 to 20 per cent larger than the females, at between 77 and 96 cm (30 to 38 inches) long, and weighing up to 70 kg (154 lbs) – about the same weight as a small adult man. The large weight-to-size ratio indicates the chimpanzee's great strength – they are far more powerful than you might expect from their small stature.

When he observed the big males in David's community, John Brown was impressed by their strength. "The sheer mass of the larger males never failed to surprise me. Fortunately, they never focused any of their power and aggression on us – we were, to all intents and purposes, invisible to them – but when a big male walked past, you could genuinely feel the earth move."

In appearance, chimps are so familiar as to be unmistakable: much smaller and less bulky in build than either species of gorilla, with mostly dark blackish hair streaked with grey covering the whole of their body; and hairless, flattish faces, with eyes set deep beneath their large, deeply ridged brow. They have wide mouths with thick lips, and large, rounded ears.

TODAY, FOUR SUBSPECIES of chimpanzees can be found across a broad swathe of Africa either side of the Equator. In West Africa, one subspecies ranges from Senegal in the west, through Mali, Sierra Leone, Liberia and the Ivory Coast to Ghana. Then there is a gap in their distribution until two more subspecies, which range from Nigeria and Cameroon to Congo and Angola. A fourth subspecies lives further east: from South Sudan in the north, through Uganda, Rwanda and Burundi to Tanzania.

Across their range, chimpanzees favour forested habitat; but are very wide-ranging in their choice of the exact type: from montane and sub-montane forests – to an altitude of 2,800 metres (over 9,000 feet) above sea level – in the highlands, to swamp and dry forests in the lowlands. If they survive their early years, when they are most vulnerable, chimpanzees usually live to 30 or more years old, with some individuals perhaps reaching 50.

The dry season begins

One of the reasons chimpanzees are so successful and widespread compared with the other great apes is that, unlike their mostly vegetarian cousins, chimps are omnivorous. And during the rainy season, which takes place here from May through to September, food is plentiful.

Throughout their whole range, chimps can and do feed on a very wide variety of fruits (more than 300 different kinds have been recorded), vegetable matter, fungi, eggs, ants, honey, honeybee larvae, and the flesh of any mammals they catch – mainly other, smaller primates such as monkeys. Their catholic diet enables them to take advantage of a far wider range of habitats than the gorillas or orang-utans, which are almost entirely vegetarian and so confined to forest habitats.

The diet of any one community of chimps may differ from that of its neighbours, and can also change considerably through the year, depending on the availability of different foodstuffs. However, fruit is always at its heart. For David's group, the rainy season always provides more than enough fruit for them to eat. During this period, the group are usually more dispersed than at any other time, as food can be found more or less anywhere.

But when the dry season arrived, and food started to become far scarcer and less easy to find, the behaviour of the whole group began to change.

Chimps are also capable of using tools: in this case, a stick which they probe into a termites' nest to catch these social insects.

A group of chimps in the Ivory Coast, using a stone as a hammer to crush the outer casings of fruits.

Instead of feeding all the way across their large territory, in small, dispersed groups, they started to come together, the whole group gathering around huge termite mounds.

As the land dried out, and temperatures rose, these mounds became magnets for the whole group, eager to supplement their meagre diet by feeding on high-protein food: termites, which might make up over a quarter of their daily intake.

Because the coming of the dry season meant there was less food and water available for the chimps, tensions between the males soon began to rise. And when a female then came into oestrus, the tension mounted even further, as the males began to become more and more aggressive, as Mark MacEwen noted.

"You could see that David was beginning to feel the pressure. He had a subtle tell-tale indicator that he was becoming stressed: flicking his toes. When he did that, I'd often see Luthor sizing him up and picking those moments to rush through the group and throw large boulders around. David would frequently respond to this by bipedally displaying, shaking large trees and throwing boulders himself."

PERHAPS THE MOST unusual feature of chimpanzee feeding behaviour is their ability to use tools to obtain a wide range of different foods, a behaviour that was first spotted by primatologist Jane Goodall in the early 1960s.

Until then, it had been assumed that human beings were the only animals clever and resourceful enough to make and use tools; indeed, reports of chimps cracking open nuts with rocks or pushing twigs inside termite nests were routinely dismissed as fantasy or fiction.

Since then, scientists have had to completely reassess those views, especially because several other species – including various different kinds of crow – have also been shown to have the capability to make and use tools.

For chimps, the simplest use of tools involves the animal picking up a suitably pointed stick, shredding the end with its teeth to create a rough surface on which the insects will grip, and then carefully inserting it into a termite mound so that the insects grab on and can be pulled out. But some chimps go one step further: they have been observed to find a smaller stick and fashion it into a hook, in order to dig out recalcitrant termites from their hiding places. Sometimes they even use tools sequentially: breaking open a termite mound with one tool, and then fishing out the contents with another. They also use tools as a way of getting ants out of a nest without being bitten.

Other kinds of tool used by chimps include the use of a combined set-up, in which a rock or stone is used as a hammer to break open the hard, outer casings of fruits, using a large stone as an anvil. Once a group has discovered a suitable stone, they will use it again and again; perhaps for decades. They will also use objects such as rocks, sticks and stones not as tools, but as missiles, especially in fights between rivals or with neighbouring groups.

These abilities are passed down the generations: young chimps will often sit and watch their mother as she attempts to fish for termites.

What is really fascinating about tool use in chimps is how it differs between different populations and different groups, so that each community of chimps will approach and solve a problem in slightly different ways. This has led to the idea that, like humans, chimps can be said to have different 'cultures'.

GATHERING AROUND THE termite mounds to feed – sometimes for hours at a time – was vital if David's group were to survive the rigours of the long dry season, which would continue until the following May. But as the chimps came together, this also posed a major threat to David's position as alpha male.

Now old friends, and enemies, were feeding side by side; and there were two ambitious rivals David had to keep a close eye on. Jumkin was at this stage number two in the rankings observed by the scientists studying the chimps, so in theory posed the greatest threat to David's rule. He was a feisty warrior, who was clearly a key contender for the top spot.

But David also had to watch out for Luthor. Although he was at that stage lower down the rankings than either David or Jumkin, he was still a major threat to the stability of the group.

Chimps groom
themselves regularly to
remove parasites from
their fur.

It was time for David to forge new alliances. Unless he could get back-up and support from at least one other male, to fend off the competition, he would be in a perilous position. Looking around the group for a new ally, his eye settled on KL, a mature male with lots of experience. Being older, KL had gone past the age when he was likely to challenge David for the position of alpha male; for him it made more sense to be subservient to David, and so receive some of David's patronage.

David approached KL and began grooming him, using his hands to comb through KL's fur: a sure sign that he would like his support. For KL, responding positively to David was a clever tactical move: by making an alliance with David, KL would have a better chance of mating with one of the females, and siring his own young. But only time would tell if this new alliance would pay off, and if David and KL would be strong enough to fend off any power grab by a younger male.

Watching David closely, John Brown began to draw conclusions about his character: "I'm convinced David is an introvert, or at the very least, he needed time alone to decompress from the pressures of life. We'd often find him, alone and slightly separated from the group, fishing for termites. It seemed almost a meditative behaviour for him. He seemed to be getting almost as much nourishment out of the solitude as he was from the termites themselves."

Drought and fire

With temperatures reaching 50 degrees in the shade on some days – the pressure on David was building. By March, the landscape had turned into a dustbowl, making food even harder to find. And the combination made tempers fray amongst the males, so it was not long before trouble began.

It started without warning when David's rival, Jumkin, suddenly turned on an older female. He isolated her in an open area of scrubby forest, then stood towering over her, pulled up a branch and repeatedly thrust it down towards her as she cowered on the ground. As she tried to fend him off, he grabbed her and pushed her down again and again, before running off.

It is hard to know exactly why he did this. Males do often attack younger females – when a female comes into season she is more likely to submit to mating with a male who has previously dominated her. This female, however, was well past breeding age. Perhaps it was just Jumkin's way of provoking David into a reaction by testing whether he, as alpha male, was still willing and able to protect vulnerable members of his group. Or maybe Jumkin was simply frustrated.

Whatever the reason for his aggression, Jumkin did not have to wait long for a response. David's new ally, KL, stepped in on his behalf, giving Jumkin a lesson in who was in charge. David then finished the job, in a clear expression of dominance over Jumkin.

This rapid response worked: Jumkin was quickly put back in his place – partly because he now knew he faced a strong alliance between David and KL, who was proving to be a loyal and reliable supporter. But would this be enough to stop any threats to David's power over the coming months?

The reaffirmation of David's position as alpha male came just in time. Fires had by now destroyed three-quarters of the group's territory, obliterating many areas where they would shelter or find food. And the temperatures continued to soar. With water scarcer than ever, the group gathered together in a dried-up riverbed, where they used knowledge imparted down the generations to dig wells, and find the precious, life-giving liquid beneath the parched earth. Using their powerful front limbs, they scraped away the dry earth. Soon the hollows that they dug filled with water for them to drink.

In general, chimpanzees are very adept at finding sources of water, especially during the long dry season. If digging wells is not an option, some groups of chimps will gather leaves or moss to be used as a sponge, to soak up any water hidden in hollows in trees. If they find a waterhole, they often drink by cupping their hands together or folding leaves to form a makeshift container.

During his challenge to David, Jumkin made a show of aggression (top), using a stick as part of his display (centre), and running away on his hind legs (bottom).

KL, an older chimp, was a useful ally for David.

An attempted coup

Like humans, but unlike many other mammals and almost all birds, female chimpanzees do not have a specific 'breeding season'; instead they come into oestrus each month, unless already pregnant or nursing a baby.

When they do come into season, female chimps send clear signals to any males in the vicinity that they are ready to breed – as well as emitting hormones, their genital area also becomes very swollen.

As with many creatures, the aims of males and females are very different. Males mate with as many females as they can, because that maximises their chances of their genetic line continuing down the generations. If they fail to breed, as many males do, their genes will die when they do. One of the main advantages of being an alpha male is that it gives David first right to mate with the females, maximising his chances of siring heirs, while at the same time minimising the chances of his rivals doing so – hence the increased aggression between males when females come into season.

An ovulating female only has a short window of opportunity, lasting roughly a week, to mate with as many males as she can. This leads to what scientists call 'paternity uncertainty': a male chimp can never be absolutely sure whether any new-born infants belong to him or to a rival, so he will hedge his bets by protecting them all. Despite this, infanticide has occasionally been recorded amongst some groups of chimps.

Mating behaviour is not the same throughout the female's period of oestrus. At first, when she comes into season, she will mate with as many males as she can; but as she reaches the peak of her spell of oestrus, when she is most likely to be ovulating, the dominant males become more interested, fighting off any lower-ranking animals. This can result in serious fights between rivals.

Occasionally, different mating strategies occur. A male and female may pair up, and both leave the group temporarily to mate elsewhere, on their own – known as 'consortship mating'. Alternatively, on rare occasions a female chimp may leave her own group and mate with a male or males from another community – known as 'extra-group mating'. But she must be very careful to cover her tracks; if the males from her original group realise that she has gone outside, they will not let her come back.

Because of this group's isolation, the females in David's community rarely had the chance to mate elsewhere – and the preponderance of males in the group meant they didn't really need to. With so few females, most

Luthor, David's younger rival, sometimes showed aggressive behaviour to other members of the group.

of the time only one was in season; the rest were either pregnant or nursing infants, which meant they would not become fertile again for two or three years – a much shorter period, incidentally, than forest chimps.

Yet in this group, no fewer than three of the seven females came into season at exactly the same time – a rare coincidence. This attracted the attention of all the males: whether they had the right to mate or not, every one of the 12 males in the group would try to – with or without David's permission. There had never been a more important time for him to reassert his power. But as testosterone levels reached their peak, David was being pushed to the limits of his ability to control the group. It really was every male for himself.

This was the most dangerous time for David – and an opportunity for his two closest rivals. Some chimps in the group had succeeded in killing previous

alpha males, so the danger was very real indeed. David – and his loyal ally KL – were now seriously outnumbered.

By then, Rosie and the team had learned to read the signals. "We always knew when something was about to kick off – whether just a small display or a big fight, you could feel a mood amongst the group. Witnessing a chimp in full-on fight mode is like watching a world-class boxer – their power and precision really is like nothing else I have ever seen. And the sound just blows you away."

Darkness fell, and at some moment during the night the younger males turned on David, beating him, biting him mercilessly. Given his strength and authority, joining forces with other males was the only way they could overcome him. Finally, after this coordinated and frenzied attack, they retreated, leaving him for dead.

David was left with serious injuries: a missing thumb, a gash on his leg, a punctured scrotum and lacerations on his rear end, arms and head. Rosie remembers that when they came across him the next morning he looked terribly weak and vulnerable. "The day we trekked out and found David alone, fighting for his life, I felt utterly heartbroken. He looked so vulnerable, and in such pain, yet there was absolutely nothing we could do. When he moved he was totally silent, as he tried not to alert the rest of the group to his whereabouts."

That day, however, the advantages of being the alpha male emerged. David's females came to tend his wounds, licking and cleaning them to enable them to heal more quickly. This was a remarkable display of tenderness, as they expertly worked on his injuries. It was also a touching display of loyalty to David, given that at this stage, his days as alpha male appeared numbered.

But having helped him, they and their infants had to depart, as they needed to find water. For the other males in the group, this was their opportunity to assert their authority over the others and take over his vacant position as the alpha male.

Meanwhile, David had no choice but to lie low. With his injuries, even if he could keep up with the rest of the group, as they moved onwards under the burning sun, he knew he would not get a welcome reception if he found them. When they saw him, they would attack and kill him. So, he stayed behind, totally alone.

At this point, Mark MacEwen – who spent the days afterwards watching and filming David – was convinced that his dominance over the group was finally over. "I genuinely didn't know if he would survive his injuries, as they were so severe, and he looked like he had given up when the group left him behind. I sat with him for days, and it was a very strange experience watching such a majestic powerful character at his weakest moment, wondering if he had the strength to keep going and being unable to do anything but watch. It felt much like watching a friend suffering, as I'd spent a long time with David and the group and got to know so many of their individual characters. Chimps always appear so strong and powerful, so seeing him so weak was a real shock."

At one point, a group
of rival males attacked
David, causing him
serious injuries.

108

The females tended David's wounds, but then had to leave to rejoin the main group.

An unlikely return

ABOVE

Cameraman John Brown
filming a group of chimps.

OPPOSITE

David returned to the
group and managed to
reassert his position
as alpha male.

OVERLEAF

David eventually
recovered from his
injuries and returned
to full health.

With the community in disarray, leaderless and directionless, David's junior rival made his move. This might prove to be the best opportunity he would ever have to move up the ranks, and perhaps even become the alpha male: taking advantage of the temporary power vacuum at the top, and moving in to seize control.

Luthor began by intimidating his rival males with a show of force. With David around, this would have been carefully controlled. But Luthor seemed to be totally out of control: terrorising the whole group – females and infants as well as males – with his violent and unpredictable behaviour.

Meanwhile, despite his injuries, David managed to find the strength to search for – and find – food. He, too, only had a short window of opportunity: he had to get fit enough to re-join the group.

So, several days after the original fight that had left him so badly hurt, a recovering David headed back towards the group, a journey of ten kilometres (six miles) under a blazing sun. The moment of David's meeting with Luthor was crucial. He had to look as if he was back to full health, even though his wounds were still healing. This was an all-or-nothing encounter. David terrified Luthor into submission by his sheer force of personality. As chimps do when they need to assert dominance, he made himself look bigger by raising the hair on his body, and standing up on his hind legs.

The future for chimpanzees

In Senegal, David and his group are under threat from habitat destruction
(especially by fire used to clear the ground for farming), as well as from
poaching and disturbance.

The gold-mining industry here is also growing at a rapid pace, creating
several problems for chimps across this region. The mercury which is used
to pan for gold may be contaminating the limited water sources which the
chimps rely on, miners often set fires to clear the land, roads are being built,
and large quarries, holes and tunnels are being dug. It also means that there
are far more people roaming in David's territory than there ever were before.

Chimpanzees are facing similar problems throughout their range. Even
though they (along with their cousins the bonobos) are our closest relatives,
sharing roughly 98.5 per cent of their genes with human beings, we have
not treated our cousins well. Because of us, chimps are declining rapidly,
and disappearing from many parts of their former range. The species is now
officially classified as 'Endangered', meaning that the threat of extinction
sometime in the future is very real indeed. The Western subspecies, of which
David's group are part, is now considered to be 'Critically Endangered'.

There are a number of reasons for this. Some are shared with many of
Africa's largest and most charismatic mammals: the loss and fragmentation
of their forest habitat; poaching for bushmeat, not just to feed local people
but as a commercial trade; and hunting to take into captivity – these are all
major issues. And because chimps live a long time, and so reproduce fairly
slowly, a population takes much longer to recover from any decline than
would that of a shorter-lived species.

The rapid increase in the human population of sub-Saharan Africa,
predicted to rise by more than one billion people by 2050, will inevitably lead
to pressure on land and greater deforestation. Logging and mining industries
also conflict with the survival of chimpanzees. And infectious diseases such
as Ebola fever are a major problem for chimps, as they are for all great apes –
including humans.

Conservation organisations are doing their best to preserve the habitats
of chimpanzees, and also to safeguard them from poaching. Despite these
efforts, only one in five of the world's chimps live in protected areas and, even
where there are clear laws against killing chimps or destroying their habitat,
these are often either enforced weakly, or not at all.

The rains arrive

Just two months after the conflict between David and Luthor had ended, the storm clouds finally began to gather above the savannah and forest, and the rains began to fall. After eight months of drought, the landscape was completely transformed and, with plenty of water everywhere, the group no longer needed to spend so much time together. Nor would they need to travel as far in search of food and water: there would soon be plenty for everyone. But at the start of the wet season, the fruits on which the group would feed were not quite ripe, so to bridge this temporary gap in food resources, they turned to hunting.

Elsewhere in Africa, chimpanzees often hunt as a group, led by the alpha and other senior males, who pursue their prey through the forest until it is caught. Females will also frequently hunt; indeed, in some communities of chimps, they appear to perform the main share of hunting activities.

Although sometimes fights break out at a kill, more often than not the dominant males will then share food with the other members of the community – especially with any females that are in oestrus. Sharing may seem at first sight to be an example of purely selfless behaviour, but Rosie Thomas takes a more nuanced view of what they are doing. "They often share their spoils with other members of the group, to curry favour for later. Everything that happens

Close bonds between
members of the group are
vital for survival.

within a chimp community is a political manoeuvring of some sort or other; it's like watching the inner workings of the White House or the Houses of Parliament playing out in the wilds of Africa."

The image of chimps as comical, gentle creatures (itself of course a myth) has taken something of a blow in recent years, as wildlife films have revealed them to be highly effective predators, chasing down and killing colobus monkeys with devastating skill.

As Mark MacEwen recalls, this group occasionally went after baboons. "Watching them on a baboon hunt really brought home how powerful, intelligent and organised they were. We could see for ourselves how they cooperated, working together as a team to herd the victim towards each other. Seeing them hunt like this reminded me even more than usual of their closeness to humans."

Mark is not alone in noticing this similarity: some scientists have suggested that the evolution of human beings may have originally been driven by the need to learn to hunt cooperatively, which then led to bigger and more complex brains.

Here in Senegal, however, there are very few monkeys, so these chimps have developed a combination of hunting and tool use to catch another, much smaller primate: the galago or bushbaby. They do so by sharpening a thin stick to a point, and then using it to skewer the galagos inside their nest holes in trees, before removing them to eat.

Rosie was captivated by the chimps' skill and ingenuity. "Breaking a branch, stripping it and sharpening the tip to try and flush out galagos from tree holes is extremely clever, but to wait until the rains have fallen, as they then know that the galago will be pushed up its tree hole as it fills with water, is just amazing."

THE COMING OF the rains had allowed everyone to relax a little after months of hardship, and also bought David time to regroup and refocus. He could not let his guard down: as always, he needed to show strength in front of the others. But time had started to heal his injuries, and he was much fitter and healthier than he had been just a few weeks before.

John Brown regards David's recovery as almost miraculous. "He sustained injuries that would almost certainly have killed a human – if not through initial blood loss, then undoubtedly through subsequent infection. He recovered at the toughest time of the year in terms of environmental conditions, while coping with the pressures that come with his position of dominance. This was a testament to just how extraordinary a creature he is."

Although David was back on top, the last few weeks had taken their toll – both physically and mentally. Only now could he start to relax a little, as he recuperated by gorging on ants. But even though he was more or less back to full fitness, he still might not be able to hold on to power.

David spent a lot of time grooming his fellow males, removing parasites and dead skin, and, in turn, being groomed by them. Every time he did so, he was strengthening the bond between them – in human terms, David was on a charm offensive. Next time another female came into season, the males would lose their hard-won discipline again, and he needed to have total authority over them. If there were to be another insurrection, he might not be lucky a second time.

As the weeks went by, he surrounded himself with some of the older males, building a band of allies. In return, they gained the advantages that came with joining forces with such a strong, resilient and experienced alpha male as David. The team referred to David's new allies as "Granddad's Army" – a highly experienced group of chimps, well able to support David.

Then came the acid test. Another female came into season, piquing the interest of the males once again. Had David done enough to be able to assert his position and get the lion's share of mating opportunities?

Last time David had been very vulnerable, but this time he was much better prepared, with a team of allies around him – including his loyal friend KL – to maintain discipline and put off any potential rivals.

The young chimp with this adult male is now rapidly gaining independence, so is no longer being carried by its mother.

Some time after they had mated, when the female was ready to give birth, she disappeared, on what the scientist studying these chimps calls 'maternity leave': it is much safer for any female to give birth away from the group, and return a few weeks afterwards.

So, roughly nine months after he and the female had mated, the fruits of David's efforts appeared, in the form of a new male infant – David's son and heir. His mother will carry him for the next year or so, as he clings to her chest and hangs upside down beneath her, or holds on for dear life as she swings acrobatically through the trees.

Then, as he grows bigger and stronger, he will start to travel on her back. He will also share her nest at night, and do so until she gives birth to another infant, while he will feed on her milk until he is two or three years old. But long before that he must start to become independent, playing with other youngsters in order to learn the social skills he will need – especially if he is to become the alpha male.

CHAPTER THREE

Painted Wolves

MANA POOLS NATIONAL PARK is one of a kind. It's a distinctive landscape of floodplains, forests, islands, sandbanks, pools, pans and rivers on the south bank of the Zambezi River in northern Zimbabwe. Its name is derived from *Mana*, meaning 'four' in Shona, a reference to the four permanent pools that form in the meandering Lower Zambezi Valley.

Life here, as in all parts of the world, is governed by the rhythm of its seasons, in this case wet or dry. In the rainy season the floodplain turns into a broad expanse of pools and pans, flanked by lush woodlands of mahogany, wild figs, ebonies, baobabs and acacias. During the dry season the Zambezi, its four enduring pools and a few of its tributaries are the only water for miles around, so the place is a magnet for thirsty wildlife.

Hippopotamuses hog the shorelines, while elephants, zebra and large herds of Cape buffalo move into the area just to drink. They join floodplain residents, including impala and waterbuck at the water's edge, while crocodiles, lions, hyenas and painted wolves are waiting, ready to intercept them.

It is amongst these riches that lived Tait – super-mum and alpha female, the undisputed leader of a truly extraordinary dynasty and the principal player in this true-life saga.

Tait's pack

Tait (right), leads her pack across the dividing river into the Lion Pridelands.

Tait was a painted wolf, an endangered species of pack-hunting wild dog that lives in East and Southern Africa. It's also known as the African or Cape hunting dog, but a literal English translation of its scientific name *Lycaon pictus* is 'painted wolf', a name that many people who have come to know it well actually prefer. It's well named, for the painted wolf plays a similar role as the grey wolf in other parts of the world.

The painted wolf is about the size of a labrador, with long, slender but muscular legs, a thick neck, a broad head and powerful jaws, so it resembles a small wolf more than a dog. Its large, saucer-shaped ears have an abundance of muscles so they can be swivelled like radar dishes, not only picking up the quietest of sounds, but, because of their size, also helping to dissipate heat. The painted wolf's coat, which is of bristle hair with little or no underfur, has a striking pattern, as if an abstract impressionist has daubed it with black, brown and white, and each painted wolf has its own unique pattern. Colours are different on either side of the body, which meant that to identify, track and film individual animals, the production team had to be able to recognise 182 flanks on the sides of 91 painted wolves in three different packs!

Painted wolf packs are headed by an alpha female and alpha male. Generally, they are the only pair in the pack who breed, and they are supported by subordinates who may or may not be related to the alpha-pair. In Tait's pack, she was the alpha female and her alpha male was Jed. He sired her babies, looked after them religiously, along with the rest of the pack, but he was very much consort to the queen. Tait decided what the pack would do: where and when they would hunt, what they would catch, and when they would rest.

Many of their offspring hunted with them, forming a well-rehearsed attack pack, with the stamina and determination that saw them regularly run down and

catch impala, a species of antelope three times their size, that were amazingly agile and extremely fast.

Even though the pack appeared to work as one, each member was a character in their own right, and some had their foibles. Tammy was the boisterous one. She was Tait's youngest daughter, and, if there were fun to be had, Tammy would be causing the mischief. Tait Junior was at the other end of the spectrum. She was the social climber and seemed, at least to experts observing the pack, like a leader in waiting. Then there was Lunar, who had a thing about hippos. Every time the pack passed the river, a pool or even the smallest pan, she would stop and gaze at them, somewhat vacantly, but she was watching the wrong animal! The rest of Tait's family would be 100 per cent focused on crocodiles or the slightest hint of them, a real threat. On reaching the water's edge they would line up, stand and stare into the water, but they would seldom cross. Even if prey were on the other side for the taking, they would not enter the water. To them, crocodiles were something to fear, sinister beasts that painted wolves were wise enough to leave well alone.

The crocodiles that live in the valley are Nile crocodiles, the world's second largest species. Old-timers might grow to five metres (16 feet) long, and they have such large jaws they are quite capable of swallowing a painted wolf whole. They can turn up just about anywhere. During the wet season, when Mana's many streams, pools and pans fill with water, crocodiles move out of the river and into the many pools dotted around the landscape. At this time of the year every pool could have its own crocodile, and painted wolves, like any animal, need to drink, so Tait's pack had to be wary of these giant primeval reptiles. So they gaze at crocodiles. Director and producer Nick Lyon was surprised at how much of their day was occupied in watching them.

"These dogs spend a lot of time staring at crocodiles," he recalls. "It's one of their biggest pastimes. It's hard not to underestimate quite how much of their day is spent staring into the water, just to see where the crocodiles are. Even if they had no reason to cross it, they would stare and stare and stare. They're fixated on crocodiles." By the time Nick and his production team came to live with the crocodile watchers of Mana Pools, Tait was by far the oldest and wisest alpha female. She had been the undisputed queen for close to six years. She ruled the floodplain uncontested, but she was showing signs of wear and tear. At nine-and-a-half years old – considered very old for a painted wolf – her strength, speed and influence were beginning to wane. Nevertheless, Tait was able to use her experience to make up for the deficiencies of old age, and she still controlled the best hunting and denning territory in the area ... but all that was about to change. A storm was brewing, and it had nothing to do with the weather.

Coup d'état

Tait's first setback was that she lost her alpha male Jed and several of her pups, which were killed by lions. Generally, these larger predators did not plague Tait's territory but, just occasionally, especially in the dry season when animals were heading for the river, her pack had to watch their backs.

At the time of Jed's death, it was thought that all of the litter of that year, bar the boisterous Tammy, had been lost, but one of Tait's boys had survived. He must have been lost when he was only four months old, yet somehow he had managed to cross the Pridelands on his own and reached Janet's territory in the east. Janet adopted him, and he helped to care for his big sister's puppies.

Fortunately for Tait, new male Ox came along. From where nobody knows for sure, probably somewhere to the west, but she and her daughters accepted him, so he stepped into Jed's shoes as alpha male. Having come from outside the territory, he had little knowledge of the layout at Mana Pools, and had to learn the ropes; but no sooner had he arrived, things took a turn for the worse.

On one fateful morning, Tait and her pack were resting; then, one by one, they became very much alert. There was a familiar scent on the wind, but a smell that shouldn't have been there. Blacktip and her pack had wasted no time. As they approached, they stopped momentarily. Ears were erect. Then they started to run, bearing down on the territory holders before they had to time to react. Nervous antelope stood around, confused. As the painted wolves went streaming past, they somehow knew that they were not the targets that day; instead, they were spectators to a royal coup.

ABOVE

After Tait limped through
the water, her pack
hurriedly follow her,
nervous of crocs. They are
now entering the
Lion Pridelands.

OPPOSITE

For painted wolves, lions
are the greatest natural
threat. In their time
working with the packs,
the production team
estimated at least 10
deaths by lion.

The clash was noisy and bloody, but Blacktip's pack outnumbered the opposition more than two to one. Tait had only one option: to withdraw. Her daughter didn't give chase. She looked on as Tait disappeared into the scrub. She had won the space that she needed for her large family, but at what cost? Tait's team had given as good as they got. Bloody necks, cheeks and flanks betrayed the price Blacktip's pack had to pay for the takeover.

Tait headed east. It was the only direction she could go. At first she was in the relative safety of the eastern part of her old territory, but she couldn't stop there. Blacktip would find her. She reached a shallow river that marked her old boundary. She paused, but she knew she had to go on, even though across the stream was the Lion Pridelands.

Feeling her age, Tait could only limp slowly through the shallow river but she still led the way and her pack splashed across quickly, for fear that there were crocodiles lurking. A back marker stopped, turned and checked they were not being followed, before it raced on. This was not a place to loiter. They were entering lands of which they had very little knowledge, although Tait had been here many years before. She had once tried to settle there, but her pups had been victims of lion attacks from the very same lions that lived here now. Somehow, she had to try again. It was clearly not a place to make a permanent home, but she had to make do with this temporary refuge. Tait had to bide her time before her pack could attempt to return to her old territory, but only if she could keep them safe. These were dangerous times.

'It is all about getting inside the heads of the dogs, and finding out what is affecting them on a daily basis: what their challenges are.'

Two-year challenge

ABOVE

Barrie Britton and Tait's pack survey the 'adrenaline grass'. The guides call it this because in the Lion Pridelands you can very easily find yourself face to face with a resting lion when crossing it. Not good for a painted wolf or a camera operator.

Following events as they unfolded, the production team was faced with a huge challenge. For about three-quarters of the year, painted wolves are on the move in their constant search for food. During the remaining quarter, they are denning, so are anchored to a confined area. However, it meant that, for most of the time the team was in the field, they had to travel considerable distances, scouring a pack's home range of 400–600 square km (155–230 square miles), and even then the animals might be very hard to locate in dense and impenetrable vegetation.

Although people had filmed painted wolves before, they had mostly focused on short, selective sequences showing them hunting, usually at the easiest times of year to find them. Aside from Hugo van Lawick and Jane

ABOVE

Producer–cameraman Nick Lyon sometimes wondered who was tracking whom. The packs had become so accustomed to the team travelling with them that they would often join the crew themselves, especially in the quiet wet season months.

Goodall, when making their memorable film *Solo: the story of an African wild dog* (1973), only scientists had followed groups of painted wolves over long periods. The *Dynasties* film crews were to be with them, night and day for even longer – for 669 days, to be precise – to tell the detailed story of the everyday lives of painted wolves, in every season, in all weather conditions, both by day and night. As producer and director Nick Lyon explained, his approach required a different kind of filmmaking – more immersive, and requiring enormous dedication from all involved.

"It is all about getting inside the heads of the dogs, and finding out what is affecting them on a daily basis: what their big challenges are."

Nick, of course, had challenges of his own. Although the team was fully aware of the pioneering nature of their venture, nothing had prepared them for just how difficult it might be. The initial problem was simply finding the painted

Cameraman Warwick Sloss (left) and elephant whisperer Nick Murray (right) up close and personal with Grumpy and Boswell. Boswell is one of two old bulls who can fully rear up on his hind legs, and Warwick was treated to a ringside seat.

OPPOSITE, BOTTOM

Thermal camera operator Justine Evans and guide Simeon Josia hand over to guide Henry Bandure and cameraman Mark Yates for the day shift. Meanwhile Tammy's pack are fully occupied with their croc-watching hobby.

wolf packs in the first place. Mana Pools National Park stretches over an area of more than 6,760 square km (2,600 square miles) – roughly the same size as Devon – but it was not just the size of the park that was daunting.

Changes in the weather influenced where and what the painted wolves were doing. The arrival of the rains, for example, was a mixed blessing. The prey animals were far more elusive and spread out, as water could be found almost anywhere, enabling them to disperse across a wide area. During the wet season, the open forests near the river were criss-crossed with rivers and small streams, any one of which could have been harbouring a crocodile. On the very first recce to the area, even before the team had started filming, one of the team witnessed one of Blacktip's yearlings being taken and killed by a crocodile – no wonder they stood and stared at them!

The result was that Mana's painted wolves understandably feared water, and so at this time of year they moved inland, into thicker, more forested places. This, in turn, meant that the wolves' movements were even harder to follow than usual.

"We really didn't know what to expect," explained cameraman Warwick Sloss. "The first shoot we did... it was in the wet season. They can be anywhere within the park and they can be in deep cover. I don't know how many miles we covered, but it was a lot. It was certainly many, many hours of solid driving, looking for footprints."

The team were only able to find both packs with the expert help of local specialists, including two men who know the painted wolves and the areas where they roam better than anyone: Nick Murray and Henry Bandure. Henry's tracking skills are second-to-none, as one memorable incident showed.

"We were driving along and Henry was peering out of the side of the door, narrating what was happening by reading the tracks," Nick Lyon recalled. "One day he said, 'I think they've picked up seven hyenas,' and twenty minutes later down the road, we bumped into the pack and, sure enough, there were seven hyenas."

Indeed, every morning, well before daybreak, members of the team would set off, aware that even if they did know where the painted wolves had been the night before, the pack might have travelled far away from them through the night. Yet they did in fact achieve a remarkably high strike-rate on their first trip: sixteen encounters in a month, and this included the most important one of all: the clash between Blacktip and Tait, which would set in motion a series of events that almost brought the whole painted wolf dynasty to an end.

The hunt

With Tait in exile, Blacktip and her enormous pack – by far the largest in the area – infiltrated all of her former territory. Her pack's scent marks served to redraw old boundaries, and it was time to enjoy the bounty of her mother's lands. Impala were plentiful, and Blacktip's hunting skills were textbook perfect.

Hunts are generally at dawn and at dusk and, early one morning at Mana Pools, Blacktip's painted wolves stirred and began to socialise. Sleepiness gave way to a raised state of alertness, with painted wolves interacting in a high-energy greeting ceremony – ears flattened, lowered forequarters and tail curved over the back, along with sniffing, licking, twittering and ducking under each other … and then they were ready to go.

The hunt itself started with the painted wolves ambling along, the pack spread out in an untidy line. Impala bucks looked on. Each was elegant with chestnut uppers, pale undersides and a pair of magnificent lyre-shaped horns, but they were all nervous. It was as if they were checking out which one of them might be the victim that day. The does stopped feeding and looked up, suddenly wary. While they looked, they stopped chewing, a super-alert state of vigilance for impalas. They were visibly worried. Then the pack closed ranks. Two or three already had their heads down in stealth mode, but Blacktip at the front had her ears erect. She stopped. The others gathered behind her, and they all stood stock still, staring into the distance. An impala doe had caught her eye. Something about the way it moved said it had a weakness, maybe imperceptible to us but the pack knew it.

The chosen one moved away, hiding amongst the trees, a bevy of bucks close behind her. One leapt into the air, kicking up his back legs, as if to say "Don't bother chasing me, I'm fit!" Another barked in alarm, and then turned away, revealing the distinctive M-shaped pattern on its backside. The pack, meanwhile, crept forwards steadily, heads down, muzzles thrust forward and ears pinned back against the sides of the head. Only Blacktip raised her head occasionally, thrusting her ears forward and checking on the whereabouts of the target. Other impala looked on, sniffed ... snorted ... not sure what they should do.

The pack began to split into three and fours, each subgroup in line abreast. For a short while they disappeared from view into a gully bordered by long grass, emerging like a small army in battle formation.

' The race was only a few minutes long, but it seemed like an eternity. '

BELOW

Waterbuck get caught up in the excitement of a stampede of impala as Blacktip's pack hunt through Tait's old territory.

Suddenly, an impala bolted. She had unknowingly pulled the trigger. The painted wolves were off in an instant. They ignored antelope on the margins. Blacktip had the target in view. She knew which one they could catch. The race was only a few minutes long, but it seemed like an eternity, impala jumping in all directions, some painted wolves in pursuit, but Blacktip and the fastest of her pack were unwavering.

Healthy painted wolves like Blacktip are legendary for their ability to run long and hard in pursuit of prey. They can race along at 65–70 kph (40–45 mph), with a top speed matching that of the impala. The painted wolves, however, can keep going for longer. The stamina is down partly to their physique: they are slimmer and lighter than other large dogs, with long, slender legs. They also have special adaptations: the bones in the foreleg – the radius and ulna – are locked together, the wrist bones are fused, and they have four toes on each foot (compared to the domestic dog with five), features that enable them to run further and faster for longer than many other hunters. All this brought Blacktip within spitting distance of her victim, but it was not going to give up that easily.

The impala they were chasing are also extraordinary athletes. Aside from the speed, they can leap to a height of 3 metres (10 feet) over vegetation or anything else in their path, and cover a length of 10 metres (30 feet) horizontally in a single bound, but they haven't the endurance and tenacity of a painted wolf.

Moments after the grab. Tait and the others make quick work of an impala. They will gulp down what they can immediately before running back to call in the pack. This is not out of greed, but to protect from hyena theft.

Blacktip and another member of her pack were on the heels of their target. They tried to make a grab, but the antelope put on a sudden spurt. Blacktip drew alongside, nipping and tearing at the victim's flanks, the other painted wolf just behind; then the pack leader lunged and grasped a back leg and the impala was down. The hunt was over.

It was once widely believed that painted wolves hunt in a coordinated way, with each member of the pack working with the others. So, for example, when the frontrunner in a hunt gets tired, one of those behind it takes the lead, much like human runners do in a relay race. More recent observations question this interpretation. Prey tends to zigzag or run in a wide arc, so a backmarker can cut across the curve, turning from a follower into a leader, hence the misinterpretation. It's suggested that hunting behaviour is more random, with each pack member acting as an individual, though all members of the pack have the same goal.

Hunts can last all morning, but pursuits can be short, such as a two- or three-minute dash during which the prey would either escape or be brought down. The pack could also target several animals at roughly the same time or in quick succession. One morning, the production team saw Blacktip's pack take three baboons before sunrise and then three impala after the sun came up. More generally, when the hunting was good, they caught at least one impala twice a day.

In the dust of the late dry
season, Blacktip's pack
feed on an impala, their
sixth kill of that day.

The kill rate of painted wolves is said to be high – about 70–80 per cent, compared to 20–30 per cent for a pride of lions. These figures, however, refer mainly to packs in other parts of Africa that are hunting larger, slower prey, such as wildebeest. In Mana Pools, they chase impala so the statistic is skewed. Prey often gets away the first time, and maybe even the second, but packs like Blacktip's just keep on trying, using a strategy based on an endurance sprint, until an animal is finally brought down.

The painted wolves' shearing carnassial teeth make short work of butchering the carcass; in fact, a special 'trenchant heel' on their lower carnassials – a single blade-like cusp – gives them a greater ability to tear meat, so prey is eaten extremely rapidly. An impala is stripped of meat and soft parts in minutes, leaving the skin, head and skeleton. They rarely scavenge, although the film crew witnessed painted wolves feeding on a zebra carcass which had probably been killed previously, and also a buffalo that had died while giving birth, but that was only twice out of the 1,500 hunts they witnessed.

A change of diet

And so it went on, a daily routine of hunting, eating and resting during those halcyon days of lush vegetation and plentiful game, but the wet season was giving way to the dry, and the searing heat was transforming the landscape.

The Zambezi has the Kariba Dam upstream, which ensures the water is flowing all year, so come October all the animals in the region move onto the floodplain. It means painted wolves have far more encounters with lions and hyenas at this time of year, resulting in a peak in mortality. Away from the river, water becomes scarce, and the extreme dry changes the ability of the painted wolves to hunt effectively. It happens every year, but when the film crew were there, Blacktip surprised everyone with an unexpected solution.

Elephant footprints, which had been made in soft mud at the end of the wet, bake as hard as concrete during the dry. It was one thing for the pack to pick their way slowly through this obstacle course of shallow craters, but quite another to do it at full speed. A misplaced foot could end in disaster – a broken leg. The impala had the advantage, their enormous leaps giving them the edge amongst the potholes. Blacktip had to rethink her hunting strategy, and what she came up with was something nobody had ever seen before. During some hunts, she would switch from impala to baboons.

The baboons in Mana Pools are fearsome characters, as the pack came to find out for itself. A large male chacma baboon is roughly twice the size of a painted wolf, immensely strong and armed with formidable canine teeth

OPPOSITE, TOP

Baboon for breakfast. Blacktip's pups were already eating meat by four weeks old. Here at six to seven weeks old, the pack have started to bring back body parts to teach the puppies the grown-up skills of pulling and tugging, skills that will eventually allow them to bring down their own prey.

OPPOSITE, BOTTOM

Tammy's pack are sleeping, but her hungry pups spy impala on the horizon.

that can rip into flesh. They defend their large mixed gender troops with vigour. However, while passing the night safely up in the trees at the heart of their territory, in so-called 'sleeping groves', they spend most of the day foraging on the woodland floor. Down there they are vulnerable. It was here that the pack would intercept them, and they had the advantage of numbers.

At the start of the first hunt, the painted wolves approached silently. This was a venture very much into the unknown. A phalanx of about a dozen pushed forward, their heads low and ears back. Then there was a loud bark. They had been spotted. The pack spread out and broke into a run. The baboons panicked. This was something new to them too. They ran in all directions, many pursued by a painted wolf or pairs of painted wolves. Their urgent alarm calls echoed through the forest, and the males in the troop were able to turn the tables on the pack. The pursuers suddenly became the pursued, but the end result was that while some of the painted wolves were keeping the large male baboons busy, the rest could seek out the smaller and more vulnerable members of the troop. Even so, a catch did not necessarily mean a kill. The male baboons were on them like a shot, and the painted wolves were forced to release their hard-fought prize, but with Blacktip's pack almost everywhere, the male baboons lost control. The hunt was chaotic at first, but the painted wolves were learning by trial and error how to hunt in a way nobody has ever witnessed before.

The painted wolves did not get off lightly. Blood was spilled, and several of the pack had open wounds, no doubt inflicted by those long canine teeth of the male baboons. It resulted, however, in an extraordinary demonstration of pack unity. The healthy gathered around the stricken pack members and licked their wounds. They healed incredibly quickly, and it was not long before they were up and running again.

What was remarkable about this partial change was that a pack as large as Blacktip's could risk trying something new and switch prey. The advantage was that for some of the time it was safer to go for baboons than impala. A bite was less debilitating than to break a leg, and, with practice, the pack became so accomplished at hunting the animals that they targeted the large males, which meant more meat per hunt. From then on, Blacktip's pack specialised in hunting the two species – impala and baboons – 50:50 in the dry season and 90:10 in the wet season in favour of impalas.

The dry season marked another change in the pack's daily routine. They had to abandon their more nomadic lifestyle and settle down for three months. It was the most important date in their calendar, when the next generation appeared.

When Tait had been here, she occupied old aardvark burrows as birthing dens, the same three in the same order each year. Here, the pups were safe from predators, such as lions and hyenas, but the takeover meant that the denning sites were vacant. Blacktip moved in, occupying a burrow about a kilometre from Tait's old dens.

The seasonal change in Mana Pools is very pronounced. The lush green undergrowth of the floodplain is quickly devoured by the impala, waterbuck, kudu, eland and elephant when the rains end, and within a month or two it is transformed into a dusty brown open forest.

MIDDLE

A nervous look as Blacktip's pack turn their sights on a new prey.

BOTTOM

Hornet, Blacktip's alpha male, is third in this perfect line up of synchronised stalking. When the painted wolves approach their prey, this low-slung head position and swung-back ears show they mean business.

TOP

Adult male baboons are almost twice the weight of a painted wolf, making them an intimidating adversary – but a welcome source of food with puppies to feed.

MIDDLE

Despite a valiant attempt, the sheer numbers of Blacktip's pack overwhelm the baboons and they capture a youngster.

BOTTOM

A baboon looks on in dismay as Blacktip's pack dispatch one of its troop members.

In her battle with her mother, Blacktip had won more than hunting grounds. She had secured a safe place to raise pups, and, not long after the invasion, she gave birth to five. While mother and pups spent the first three weeks in the den, by the time they were six weeks old Blacktip was out hunting again, although being anchored to the den meant that the pack had to travel increasingly greater distances from the den site in order to find suitable prey.

The pups, meanwhile, had long spells above ground. One or two members of the pack guarded them. They would all wait impatiently for the others to return with the promise of food, but one morning they were confronted by much more than they had bargained for. They were greeted by loud trumpeting – elephants!

There are more than 3,000 elephants in Mana Pools National Park, one of Zimbabwe's largest concentrations, and half a dozen of them were right outside the den, but the pups' new neighbours were only interested in food. They had been attracted to the fruits of the apple ring acacia, a favourite elephant delicacy and a vital food source during the dry season. Fruits in the topmost branches were out of reach for most other ground-based animals, so were there for the taking, except that even animals as big as elephants had trouble reaching them. The solution? One of the huge pachyderms reared up, and, balancing on its back legs, it stretched up its trunk to delicately pick the fruit. The painted wolves were simply ignored.

By this time, Blacktip had her work cut out. With the alpha male and female generally the only animals to breed, it meant Blacktip not only led the hunts, but she was also the only member of the pack who could suckle the pups. It was busy being pack leader and a nursing mother. The pups, though, were already taking solids, so every time Blacktip came back with her stomach containing about four kilograms (8 lbs) of food, she had to regurgitate their breakfast, although she was quick to sneak back a piece of meat for herself. Others in the pack also brought up their food for the pups and their babysitters. Even though the pups were not their own progeny, they were closely related, so the subordinate members were doing their bit to ensure the pack's success. Studies have shown that small packs without a band of helpers are less likely to raise pups effectively.

Food from the kill is carried back in the stomach, rather than in the mouth, because this way painted wolves are less likely to lose it to hyenas. Unlike the hyenas, which begin digesting their meal immediately, the painted wolves delay the process until they are back at the den. And, with such good hunting, Blacktip's pups grew fast and strong. The move into her mother's territory appeared to be paying off, and, with a change in the seasons, impala would once more become the staple wet season food.

Precarious life in the Pridelands

The Lion Pridelands – the most dangerous place in Mana Pools – and temporary hiding place for Tait's exiled pack.

Compared to Tait's old territory, the Lion Pridelands had less suitable prey running about and considerably more dangers. During the dry season it became prime buffalo habitat, which was great for lions but of little use to painted wolves. Nevertheless, Tait had been able to keep all members of the pack alive, and, about four months after moving in, she was about to give birth too. Finding her den was not easy.

One of the reasons that Mana had been chosen as the location to film painted wolves was that Tait was the only animal known to have frequented the same dens, which meant that if she did so again they would be able to find her more easily. This denning strategy was one of the things that had made her so successful. But, when Blacktip had driven Tait out of her territory, she had to find a new den site in the Lion Pridelands, and the team were back to square one.

At the Sand River, Tait's pack have been waiting for three weeks to greet the new arrivals: two tiny puppies soon to emerge.

After weeks trekking through thick bush infested with tsetse flies and mosquitoes, Nick Murray and Henry Bandure discovered tracks that finally led the team to the open Sand River, at the far end of the Pridelands. Here Tait had chosen a room with a view – a den that afforded wide vistas across the dry riverbed – perhaps to see approaching danger now that she was in unfamiliar terrain.

The den was an old aardvark burrow that she had taken over. It was here that she gave birth to two pups and, after three weeks in the dark, the day had come for them to emerge for the very first time. Nick Murray's long relationship with them meant that the crew were accepted by the pack, and so were able to get extraordinarily intimate footage of the relaxed and undisturbed litter.

When they were first about to surface, the rest of the pack had been resting in the midday heat. As they stirred, there was a noticeable excitement in the air. Somehow they knew what was about to happen, but first they had to perform their daily ritual. Two of the painted wolves came face to face, rubbed muzzles and dropped to their knees, rubbing, smelling and licking faces, their tails wagging enthusiastically. Then two more began. Up and down they bobbed, heads together, followed by the rest of the pack. Everyone wanted to take part. These affectionate greetings helped cement the bond between pack members, kept the pack strong.

Just as suddenly as it began, the meet and greet stopped and they all headed over to the den. Tait emerged first, blinked in the bright sunshine, and stood guard outside. The pups had been born blind and helpless, but they had developed rapidly and it was time not only to take a look at the outside world, but also to meet the relatives. A few minutes later, there they were: two little bundles of fur, unsteady on their feet, but already determined. It was a filmmaker's dream.

"We were there on the very day Tait's two tiny puppies made their first steps into the big, bright world," Nick Lyon remembers. "They were so small, and their heads so disproportionately large, that whenever they stopped walking they would teeter over their front legs, like a see-saw!"

Ox was first to go over and sniff them. It was Tait's eighth litter, but Ox's first, and he had to stand aside as the pack almost formed a queue to see, smell and fuss over the new arrivals. And, when they became too overwhelmed by the attention, they tottered back inside. The rest of the family reluctantly slunk away. Despite the temptation to stay close to the den, the dogs knew that it was safer if they took the rest of their daytime nap some distance away. Tait led by example; after all, she had done this many times before. She positioned herself so she could see the den entrance, while enjoying the breeze on her face and the shade of a large acacia tree. She also looked upon the camera crew on her doorstep as stand-in helpers. Painted wolves always leave a babysitter with the pups when they go out to hunt, and Nick Lyon found that on several occasions he was left holding the baby.

"After a while, Tait left the pups with us, and took the whole pack out hunting. One time I had to shoo them back down into the den because there was no other pack member around to do it!"

The birth of pups had been celebrated in both packs, but Tait and Blacktip had produced small litters. In a good year, they might have ten to fifteen pups each. Had the stress of the conflict affected both mothers? Nobody knows, but Tait certainly had the more worrying time. She had actually been late in denning. Painted wolves have a relatively short window of opportunity to raise a litter. Ideally an alpha female should have given birth as soon as conditions were dry enough for a den not to collapse. Then there was the pressure to get the pups mobile and up sticks and get out of there before the dry season hit the hardest. Even after weaning at about eight weeks old, Tait's pups would continue to need the den as a refuge for another month or so, but in a place overrun by lions, little was ideal.

' We were there on the very day Tait's two tiny
puppies made their first steps into the big,
bright world. '

Blacktip's pack have won Tait's territory. Now it is time to explore this lush and game-rich conquest.

For the pups, however, this was their first great adventure. A flock of helmeted guineafowl became playthings, but for a young and naive pup it must have been a mystery how they simply disappeared, flapping noisily to escape their tormentors and taking momentarily to the sky, to land on a branch in a nearby tree.

While the pups were very small, Blacktip could not wander far from the den, but now they had become mobile, she was able to roam across the full range of her territory again. The pack continued to scent mark, leaving their distinctive smell at key 'information centres' around the territory and leaving

details of who was there, how many, how strong, and even in which way they were all heading; but one day Blacktip picked up another scent. It was old, but unmistakeable – the scent marks of her mother's pack, and it gave some indication of the direction she had taken into exile.

The discovery had an extraordinary galvanising effect. Blacktip stopped in her tracks. Territorial marking ceased, and she set off in pursuit of her mother. Nobody was sure why Blacktip was doing this, but it might have been a way that, once and for all, she could ensure her mother never came back.

The terrible night when Blacktip ignored her instinct to bed down on a moonless night, and continued her pursuit. Hyenas are a much more dangerous prospect at night. And their numbers were building.

MIDDLE

Soon there were 15 hyenas facing 25 painted wolves. Blacktip launched the attack as a pre-emptive strike, hoping to force the hyenas back.

BOTTOM

Just when it appeared that Blacktip had gained the upper hand in driving back the hyenas, a pup got separated from the pack in the inky black night. Mercifully the puppy's death was quick, but it sent shockwaves through the pack.

straggler in the dark. They hadn't gone unnoticed; hyenas were tracking their every move.

Generally, hyenas were a potential hazard for painted wolves during the day, although more of a nuisance, because they would sometimes steal kills. In the dark they were a more serious threat and, on this night, increasingly more were arriving. They started to emit their loud and eerie whooping calls and, the more they called, the more hyenas pitched up, fifteen in all. They were looking for an easy meal.

Watching all this was the night crew, led by low-light and infrared camera specialist Justine Evans. With her equipment, the production team was able to observe another new behaviour. Nick Lyon had noticed the hyenas following the pack from some distance back, and the film crew's new discovery was that the hyenas had been feeding on the painted wolves' faeces. However, having ramped up their numbers they were eyeing up the painted wolves themselves.

Blacktip had made a critical mistake. She had not bedded down in a safe place for the night and it's at night that hyenas push north from the clanlands and onto the floodplain. Somehow she had to get the situation under control, for the threat to her youngsters was becoming increasingly acute. Attack was probably the best form of defence, so that's precisely what Blacktip did.

The pack was big enough to split into small groups that could gang up on individual hyenas, but they were operating in total darkness with very little coordination. It turned out to be a confusing mess, but the pack snarled loudly and nipped at the hyenas' flanks and legs, forcing them to sit on their haunches in order to protect their rear ends, and they were relentless in their attacks.

The aim was to get the hyenas to withdraw, and it seemed to be working, but, in the fracas a pup became separated from the rest of the pack, and it was taken. It was a terrible thing to have happened to the painted wolves. They had made it through the battle, but had lost a pack member, a valuable youngster. They milled about, reluctant to leave without the lost pup but, eventually, they had to go. With heads down and tails between their legs, as if in mourning, they marched in silence, a solemn procession.

Undeterred, Blacktip led the pack onwards. Eventually, they were far beyond the eastern boundary of Tait's old territory. They had entered the most dangerous place in Mana Pools – deep into the heart of the Lion Pridelands – and, having never been here before, she was leading her pack very much into the unknown. The film crew following her felt it would have made more sense for her to turn back after the hyena attack, but she didn't. She pressed on.

And into the fire

Well before dawn, two teams left camp: producer Nick Lyon and wildlife cameraman Barrie Britton travelled with tracker Henry Bandure, while a second cameraman, Warwick Sloss, travelled with tracker Nick Murray. They were deep into the Lion Pridelands trying to locate Tait. One of her pack had been fitted with a radio-collar, so, provided they could get within a couple of kilometres of it, they had a chance to find her. One of the vehicles drove through some painted wolf faeces. The smell was terrible, and Henry, who was driving, received quite a ribbing for it, but then he replied quite simply:

"Yes, but it does not smell of Tait – that is Blacktip's smell."

Just by driving through the foul-smelling dung, Henry knew that the painted wolf that produced it had been eating baboon. Tait and her pack, however, were not baboon eaters. Blacktip was. The team had found more evidence that Blacktip had left her home territory and was in a blind pursuit of Tait and her pack in the Lion Pridelands.

Tait, however, had some knowledge of the Pridelands, having been here earlier in her life and, remarkably, she had kept her pack safe from the lions for eight months, but they were on constant alert. Together, the pack members had looked after her injured daughter Wicket. Her leg was healing, but she was still limping. Life here meant that all the painted wolves were constantly on edge, and rightly so – the wind carried a warning.

It was an all-too-familiar smell. Blacktip was coming. Tait was not going to make the same mistake twice. Her pack turned tail and ran, the injured animal keeping up as best she could. It was hot; the midday temperature was close to 50 degrees. The air shimmered, and the stripes on a herd of zebra they passed seem to blur together in the heat haze; but Blacktip kept up the pursuit. She knew she was closing in. She was less than a kilometre behind Tait's pack, but after a hard night without rest, the heat was sapping their strength. Their legs were like lead.

Blacktip called a halt to rest in the shade of a large sausage tree, named for the shape of its fruit, a favourite of hippos and baboons; indeed, a baboon sat some distance away, watching the sleeping pack, along with a pair of vultures – those omens of death – but they weren't the only observers. Expert tracker Henry spotted something moving towards the pack, and cameraman Barrie Britton remembers the moment vividly.

"We suddenly noticed a lion creeping towards the dogs through some long grass. We then saw more lions following behind, and realised that the dogs had absolutely no idea they were there."

The first lioness approached from downwind, and Barrie was right. The

A daughter is lost.
Blacktip's pack look on as
more crocodiles move in
on their fallen sister.

pack was unaware of her and, for a few agonising moments, Barrie and the rest of the team thought this was it. Then, at the last moment, something woke the pack. A gruff bark sounded the alarm and they scattered, an effective way to confuse attackers. The adults growled in alarm and ran to and fro, trying to draw the lions' attention away from the pups. They were suddenly extremely vulnerable, and Blacktip could not afford to lose another. It would be nothing short of suicidal, but she was ready to fight.

Then, out of the blue, a buffalo came charging through the trees. The buffalo herd had roused the sleeping pack, and now one of them was chasing the lions. A buffalo taking on a pride of lions alone, however, was not a smart move. It twisted and fell to its knees momentarily. It was all the lions needed. As it struggled to get back up, they turned on it and pulled it down again. The pups were safe, but the buffalo had paid with its life. The pack moved on, but Blacktip's headlong pursuit of Tait was exposing the pack to grave dangers, especially lions, and lions are the commonest killers of painted wolves in Mana Pools. For the first time since the adventure began, Blacktip seemed indecisive, uncertain whether to press on.

The weather was not in their favour either. At the peak of the dry season Mana Pools' wildlife concentrates around the last of the standing water. All around, animals that rely on the river and pools get on with their everyday lives. Skimmers fly up and down, their long lower bills lowered, ready to snag a fish. Spoonbills shovel, and herons fish from whatever perch they can

BELOW

Nyami chases Ox across the side channels of the Zambezi River, as Tait's pack engage in joyful play after the retreat of Blacktip's Army.

find: submerged logs or even from the back of a hippo, the large mammals unperturbed by their unusual passengers. It appears so idyllic, but every dry season there are many large animals concentrated *under* the surface of these same pools – crocodiles!

They wait, submerged, highly attuned to the splashing of unwary animals coming to drink, including painted wolves. A crocodile attack is explosive and shocking. It launches itself from the water at the nearest painted wolf, grabs it and drags it back into the water. The rest of the pack is powerless to help, and this year, as every year, crocodiles took their toll on Mana's painted wolves, including Blacktip's pack.

Faced with such overwhelming dangers, Blacktip made the vital decision. She turned the pack around, and they all ran like the wind. They ran and ran, during the day and throughout the night. They covered close to 25 kilometres, barely stopping, all the way home. The pursuit of Tait was over. The cost had been too high.

The homecoming

ABOVE

Jemma and Tim. Post-coital head-resting seemed a common feature of the weeks when the painted wolves were pairing off and singing. This ritual is really not understood yet, but it seems like an affectionate or possessive gesture.

OPPOSITE

After Tait's death, the pack were leaderless. For many weeks Mana reverberated to the haunting calls of the pack choosing a new leader, as males and females coupled off and sang, in a ritual nobody had witnessed before.

Tait and her pack sensed that Blacktip had pulled out of the Pridelands, and their noses would soon tell them the extent of that retreat. Painted wolves can detect airborne scents from 20 km (12 miles) away, so with Blacktip's smell no longer on the wind, the pack relaxed visibly, the first time since war had broken out. It was time for them to go home too, but Tait would not be joining them.

Tait had lived a long and full life, far longer than most painted wolves, and she left a lasting legacy in these lands. Ten days later, before the pack had left for their old home territory, age and infirmity caught up with her. She was too old and too slow. She could no longer evade the lions, but Ox – her alpha male – would not leave her side. Together they died in the Lion Pridelands.

However, the story does not end there, for a new generation of leaders continues the bloodline. Life-giving rains followed the drought, and the landscape changed from dry and dusty brown to verdant green. Streams and ponds filled once more. Hard times were quickly forgotten.

It was thought that, after Tait's death, Tait Junior would take over as leader of the pack. She had already been seen scent marking with Ox. Before she died, Tait would scent mark and then Ox would mark on top of it, the sign of an alpha couple, but towards the end of Tait's life Ox also marked on top of Tait Junior's scent marks, the sign of a 'soft' takeover of the pack. When both Tait and Ox

Tammy (left) and Twiza (right), the new alphas. Tait's youngest daughter stands proud as heir to Tait's throne.

were killed, however, Tait Junior appeared to retreat into herself. What actually happened next came as another surprise.

For several weeks, Mana Pools echoed with the eerie gibbon-like calls of Tait's pack deciding who to accept as the new leader, and it was Wicket, the painted wolf who had broken her leg so badly, who rose to the rank of interim alpha female. Interestingly, Nick Murray revealed that Tait's own mother had broken her leg badly too, and it set at an odd angle. Even so, 'Broken Leg', as she was nicknamed, continued as the alpha female, running the pack and gave birth to three litters of pups … and all on three legs! Tait was one of the pups. Down the years, this pack has had extraordinary resilience in the face of adversity, and Wicket's rise to the top showed how painted wolves value knowledge and decision-making in a leader, not just physical strength.

Wicket led the pack back through the minefield that was the Lion Pridelands to the safety of Tait's wet-season hunting grounds in Zebra Vlei. At the border she noticed that it was no longer scent marked. After about ten days, a scent mark fades sufficiently for Wicket to know that its mark had not been renewed. The territory was undefended, so they moved back home. Not long after, Tait's eight daughters attracted a band of roving males who are thought to have come originally from Blacktip's pack. With the ranks of Wicket's pack reinforced, they were able to reclaim their old territory, and Wicket herself became pregnant; but there was to be one more major setback. When looking for a den site too far into the hyena clanlands, she was killed. The question that followed was who would be leader now, and the answer was another bombshell.

After a bad puppy year – with only three surviving in total from both packs – it was a welcome surprise to see 10 healthy puppies born to Blacktip in the year after the conflict with her mother.

It was Tammy – mischievous, fun-loving Tammy, Tait's youngest daughter – who emerged as the heiress to her mother's throne. Somehow Tammy had become pregnant on the sly, and, even though she was the youngest and most inexperienced member of the pack, she rose to become alpha female by default.

A few months later and hidden away in a secluded part of the forest, Tammy's first litter was born – seven pups. The births, together with the males who had recently joined, brought the pack size up to 23. In the west, Blacktip also gave birth, this time to 10 healthy pups. It meant that the two packs were roughly the same size, so neither was likely to overrun the other. A balance, and with it peace, had finally been restored to Mana Pools.

In recent years, according to the Regional Conservation Strategy for the Cheetah and African Wild Dog in Southern Africa (2015), there are about 479 packs with 4,411 painted wolves left in the wild (excluding fenced reserves) in Southern Africa, which includes the free-living painted wolves of Angola, Zambia, Mozambique, South Africa, Namibia, Botswana and Zimbabwe. Compared to the 470 packs with 4,273 painted wolves counted in 2007, numbers have shown a modest increase in roughly the same area. Another 2,000 painted wolves are thought to occur in the rest of Africa, south of the Sahara, mainly in East Africa.

Even so, painted wolves today occupy less than 17 per cent of their historical range. They are one of the world's most endangered carnivores, and humans are their biggest threat. In some places they are considered a pest and killed indiscriminately – poisoned, shot or caught in snares, although they would probably only go for livestock when desperate for food. There are no records of them deliberately attacking people. Nevertheless, about 70 per cent of painted wolves live outside protected areas, so they are at huge risk from habitat fragmentation, conflict with humans and diseases, such as rabies and canine distemper from domestic dogs. In the wild, lions are their nemesis, as both Tait and Blacktip discovered. In an area with a high population density of lions, there is usually a low density of painted wolves.

On IUCN's Red List, the painted wolf is considered 'endangered', which means that at some time in the future it is 'very likely to become extinct'.

There is, however, some degree of hope. In Zimbabwe – home to the Mana Pool packs – the number of painted wolves has increased significantly during the past quarter century and, up to the end of filming, 280 of them were known to be direct descendants of Tait. If they had all survived, this would represent a staggering 6.5 per cent of the total Southern Africa population. Tait's contribution to the survival of this species was unparalleled in painted wolf history.

Now Tait leaves behind at least three alpha female daughters – Blacktip in the west, Janet in the east and now Tammy in the middle. Her dynasty is secure again … at least, for now.

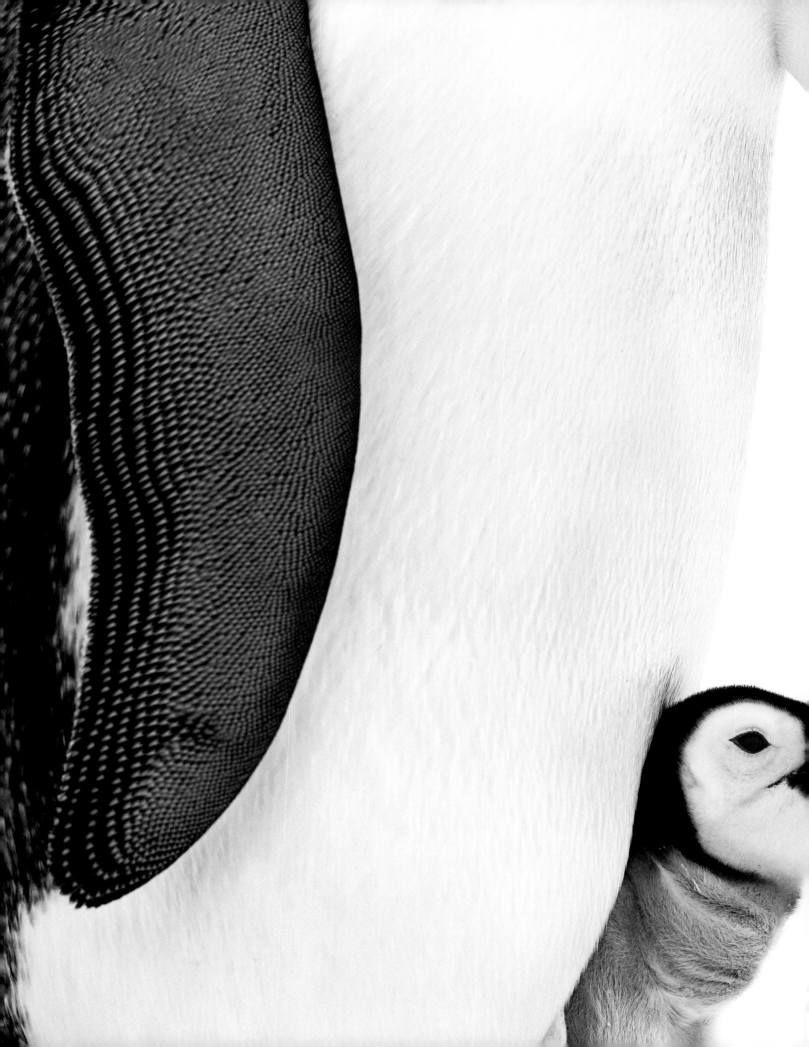

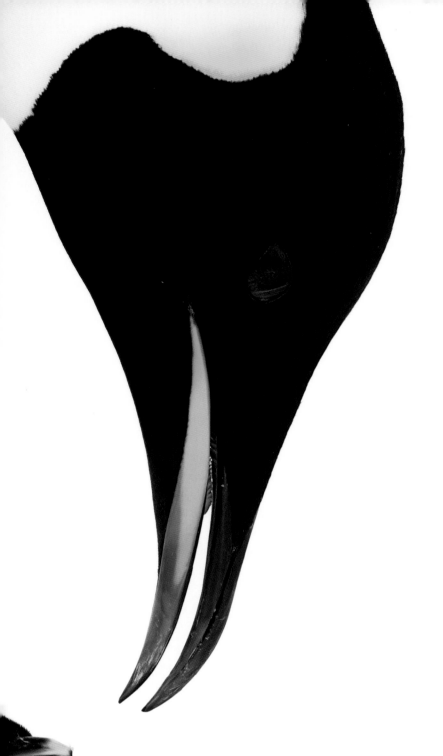

CHAPTER FOUR

Penguins

EMPEROR PENGUINS LIVE in a world of bone-chilling cold and – for several months of the year – almost permanent gloom and darkness. During the depths of the Antarctic winter the sun does not rise for weeks on end, and temperatures regularly fall to below minus 40 degrees Celsius. Where the land meets the sea, a thick layer of sea ice forms on the edge of the ocean. This is where the emperor penguins choose to breed.

The emperor penguin is aptly named: it is the largest, tallest and heaviest of the world's 17 species in its family, the world's heaviest species of seabird and, after the ostrich, cassowary, emu and rhea, the world's fifth heaviest living bird. A fully grown adult emperor can tip the scales at a maximum of 46 kg (100 lbs), and stands up to 115 cm (almost four feet) tall – truly an imposing and impressive sight.

The emperor penguin is one of two species in the genus *Aptenodytes* – meaning 'diver without wings'. Its closest relative is the slightly smaller king penguin, which breeds on various islands in the sub-Antarctic, including the Falklands and South Georgia. It is thought that these two species split off from all other penguins as long as 40 million years ago.

They are easy to identify: with a black head, tail, back and wings contrasting sharply with creamy-white underparts, shading to yellow around the upper breast and neck. Apart from the males being, on average, slightly larger than their mates, both male and female are alike, as in all penguins; and, again like all other penguins, they are unable to fly.

But the most extraordinary aspect of the emperor penguin's life is that it is the only species of bird that breeds on the Antarctic continent during the winter months, when temperatures stay way below freezing for months on end. It is only able to do so by virtue of its large size, which increases the bird's volume to surface area ratio so that it loses less heat. The emperor penguin also has several special adaptations to its body, metabolism and behaviour that make it unique – not just compared with other penguins, but amongst all the world's 10,000+ species of bird.

Yet by being perfectly adapted to such a lifestyle, the emperor penguins are able to take advantage of the fact that virtually no other bird species can survive alongside them. This means they get to exploit abundant food resources, without competition, and with few predators to put them or their offspring in danger. It's a remarkably harsh life, yet to the emperor penguin it makes perfect sense.

We often think of penguins as being birds of the frozen Antarctic; and yet other species of penguin can be found in New Zealand, Australia, South Africa and even – in the case of the Galápagos penguin – right on the Equator. Only two species – the tiny Adélie and the mighty emperor – live exclusively in and around the Antarctic continent and its surrounding oceans; rarely venturing

Nearly 11,000 emperor
penguins gather in
Atka Bay to breed.

The smaller Adélie
penguins are unable to
breed on sea ice, but
some non-breeding birds
do temporarily join the
emperor colony during
the summer.

onto any other landmass. Indeed, most emperors will never actually touch land in their lifetime, as they spend their whole existence either on ice or in the ocean.

The emperor penguin can be found from 78 degrees to 66 degrees south – the edges of the Antarctic Circle – and hardly ever strays outside these boundaries. Yet on occasion they do take a wrong turn and end up very far from home. In 2011, one young emperor became stranded on a beach on New Zealand's North Island – more than 6,000 kilometres (almost 4,000 miles) outside its usual range. Other vagrant emperors have ended up in Tierra del Fuego and the Falklands.

Despite their harsh life, emperor penguins have an average lifespan of 20 years, with some living even longer. However, just one in five of their chicks survive to their first birthday. As a result, the vast majority of birds in any one population – as many as 80 per cent – are adults.

Getting ready to breed

By and large, birds choose to breed during the spring and summer months, as there are more hours of daylight (especially the nearer they live to the North or South Poles) and therefore a more abundant and reliable supply of food for them to give to their hungry offspring.

Indeed, all the world's other 16 species of penguin do just that: starting their courtship process at the beginning of the austral spring (autumn in the northern hemisphere), and then breeding and raising their chicks during the austral summer (the northern winter), so that the youngsters are fully grown – and able to have the best chance of surviving – before temperatures begin to drop once again.

But the emperor penguin cannot follow that course, for one simple reason. Because the species is so large, if the chicks were born in the spring, they would be unable to reach their full adult size and weight during the short summer

‘ The distance the penguins need to travel may be anything between 20 and 120 km (12.5 and 75 miles) from the sea, and can take many days. ’

ABOVE

Emperor penguins take on an almost comical gait when moving over the sea ice, giving them the appearance of being poorly adapted to life on solid ground. However, when tired or crossing uneven terrain they switch to tobogganing, pushing with their clawed feet and using their beak to 'ice-axe' over obstacles.

season. The emperor is a victim of its own success: by growing so large, in order to survive such low temperatures for so long, it forgoes the ability to produce offspring rapidly.

Another reason is that the sea ice on which emperor penguins usually breed is absent during the summer months. So, the emperor has no choice but to start breeding at the beginning of winter – an incredibly risky strategy that has led to this species having one of the most bizarre and unusual breeding cycles of any living creature. But this also brings a major advantage, in that virtually all predators – and indeed all other living creatures – leave the area during the winter, so that the birds are safe at the start of the breeding season.

Having spent the summer months out on the edge of the continent, hunting for food in the incredibly productive seas offshore, the birds begin the long journey inland, where they will come together to breed in vast colonies.

Depending on the site of their colony, the distance they need to travel may be anything between 20 and 120 km (12.5 and 75 miles) from the sea, and can take many days, depending on the nature of the terrain the birds have to cross.

At the start of the
breeding season, the
colony is a cacophony
of sound, as each bird
searches for a mate.

Emperor penguins are
inquisitive birds, without
much visual variety in
their habitat, so anything
different – including the
team's skidoos – draws
their attention.

Sometimes the birds walk, shuffling along on their feet; at other times they lay down on their fronts and propel themselves along using their strong claws, like a human on a sledge.

For the film crew waiting patiently at their base camp, this was a crucial moment: would the penguins return to their usual colony, or might they either go somewhere else or be much reduced in numbers?

Camera assistant Stefan Christmann, a young German naturalist and photographer, who worked on this project as the field assistant, recalls the first time the team caught sight of the birds. "When we arrived at the edge of the ice shelf, we could hear the calls of the emperor penguins. Not just a single call but hundreds in unison, mixed into what many people might describe as a cacophony."

For Stefan, the sound was both haunting and beautiful. "The trumpet-like sounds made by the emperors are like music. I would describe them as a euphony of unique melodies!"

Assistant producer Will Lawson specialises in filming wildlife in remote locations. "I got a call to make the 10 km skidoo ride from our base down to the ice edge, as there was 'something I should see'."

When he got there, he gazed out into the distance, towards the north. "I squinted into the white haze and at first saw nothing. Then I caught sight of ghostly figures, almost human in appearance, shimmering and wobbling a kilometre or so in the distance. Once the fan of the skidoo finally shut off, the atmosphere fell silent, in a way it only ever does in Antarctica. Then, still peering north, the silence was broken by the unmistakeable cry of bird calls, far out onto the ice."

The birds were only able to return to the colony at this time of year because, with the air and sea temperatures falling rapidly, a stable layer of sea ice had begun to form, locking in stray icebergs and forming into a pattern that was likely to remain for the next eight or nine months, until summer finally came once again.

By then, the film crew had been in Antarctica, preparing to film, for nearly four months. But when Will walked down to meet his colleagues and looked toward the sea ice below them, he noticed gaping holes in its surface and ridges of ice being pushed together, showing that the ice was still not fully solid. It would be some time before they could get down at eye level with the penguins, to film their behaviour close up.

On one sunny, calm and relatively warm day – the temperature rose to minus 20 degrees Celsius, and there was no wind-chill – they did manage to get to within sight of the colony. Finally they were able to witness hundreds of birds arriving, walking across the flat icy surface of the sea ice and making their way through fields of ice ridges.

For Lindsay McCrae, the programme's sole cameraman, on whose shoulders the success or failure of the film rested, this was a moment of joy and frustration. "I'd always dreamed of seeing this spectacle, as long lines of emperors marched towards us, distorted by the haze. They seemed to appear

Cameraman Lindsay
McCrae stands on the
ice shelf looking over the
sea ice, with the emperor
colony below. This corner
of Atka Bay is the first
place to freeze and last to
melt, making it the best
place to breed.

from all directions, and covered huge distances incredibly quickly. But it was frustrating not being able to get down on the newly formed sea ice, which was still far too unsafe."

What also struck the team was how the birds knew where to come at all, given that over the previous few weeks the shifting ice had changed the landscape dramatically. It is remarkable that these birds are capable of recognising a place that has absolutely no landmarks, and can appear completely different each time they see it.

Lindsay was also intrigued by this homing ability. "How they manage to get back to the same piece of ice every year is one of the wonders of our natural world. This particular year, the sea ice had formed by freezing open water, rather than by existing ice-floes joining together, which meant that the surface was relatively flat, making the penguins' journey a lot easier."

The first birds returned by the end of March, and the vast majority of the rest arrived back by the beginning of April. "The emperor penguins have an amazing sense of timing," Stefan noted. "And now that they were coming back with their bellies filled, they looked huge and really majestic. True emperors in a beautiful coat, and more colourful than we had ever seen them before."

As each day and week passed, more and more penguins returned to the colony until ultimately there would be more than 10,000 birds present. At this

Using a modified sledge, cameraman Lindsay McCrae was able to film the males huddling against increasing winds and rapidly falling temperatures.

OPPOSITE

Snow and ice blow like grains of sand, drifting unimpeded over the landscape. It looks breathtakingly beautiful, but can also make life tricky for the penguins.

time, the sea ice between the ocean and the colony resembled an emperor penguin highway, with the birds travelling in neat long lines like people queueing – a remarkable spectacle.

Soon after they arrive, each penguin must make a crucial decision: which mate will they choose? Although emperor penguins are faithful to their chosen mate during the course of each breeding season, they do change partners from year to year – a breeding strategy scientists call 'serial monogamy'.

It may seem odd, given that the bond between male and female is so important (as we shall discover later), that they do not usually pair for life – but the reason for this is simple. Because each bird may arrive earlier or later than their former partner, and time is of the essence when it comes to getting down to breed, the early arrival cannot afford to wait for the other to arrive, and must choose from whichever potential mates are available at the time.

As he approached the colony, Will recalls, there was one main group of birds, with two smaller ones close by, all tightly packed together to retain their body heat. "The gun-metal grey of the feathers and the fact each adult was perfectly interlocked with their neighbour, gave the group an appearance that reminded me of a lizard's scales or human chainmail."

Throughout this period, the team were enjoying watching the various antics of the birds. As the sea ice was still freezing, some would take shortcuts through cracks, occasionally surfacing to breathe – looking like a shark's fin in a horror movie. Sometimes the cracks in the ice became so crowded that the whole water was filled with penguins, with little or no room for any others to get in, even if they wanted to.

How emperors adapt to life in the freezer

All penguins are adapted to a lifestyle mainly spent at sea, though this is interspersed with periods – sometimes very long periods – on land. In essence, their body shape and anatomical features are a compromise between these two very different lifestyles.

Penguins' feet are at the back of their bodies, which enables them to swim at great speed, yet are not ideal for locomotion on land, as they mean that the birds have to waddle when walking. That's also why emperors often

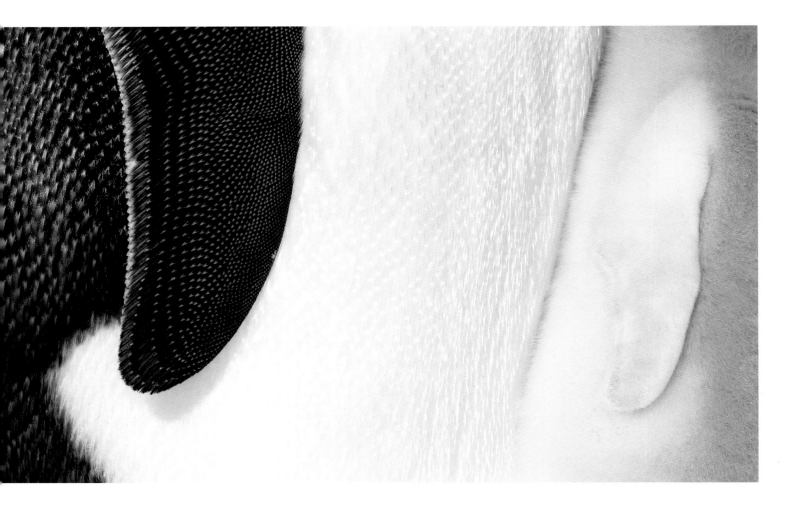

ABOVE

The emperor's counter-shaded, two-tone feathers are some of the densest found on any bird, and evolved to provide the best insulation and to be streamlined when the birds go in and out of the water.

glide over the surface of the ice on their bellies. They are surprisingly good at climbing, too: using their beak, and powerful feet and claws, to scale steep snow slopes, almost effortlessly at times.

Along with their smaller cousin the Adélie penguin, and two other species that breed in the Antarctic, Gentoo and chinstrap penguins, emperors have incredibly dense, waterproof feathers, which provide up to 90 per cent of their insulation against losing heat in these frozen environments.

Emperors also have a thick layer of blubber – fat beneath the skin – which builds up before they begin to breed, and may reach three cm (more than an inch) thick. This is far thicker than that of their cousins, especially those that breed in temperate zones where the need for insulation is less crucial.

Keeping warm does, however, come at a cost: that extra layer of fat makes emperors less mobile when on the ice, though it doesn't seem to affect them when under the water. They are also able to raise their feathers, enabling them to trap a layer of air between the surface and the skin, another way of avoiding heat loss. On several occasions Lindsay recalls seeing this for himself: "The feathers on their back would stand erect, and then – like a Mexican wave – slowly flatten from head to toe."

Another unique adaptation is that emperor penguins are able to regulate their core body temperature across a very wide range of air temperatures. Thus

they can cope in very cold conditions by moving around, shivering and breaking down their body fats to produce more internal heat. In unusually warm spells – which occasionally do occur in summer on the edge of the Antarctic continent – they raise their wings to allow them to lose heat more rapidly.

When emperor penguins enter the water, they relax the muscles that hold their feathers erect on land, making their bodies far more streamlined, and enabling them to reach their typical maximum speed of 10.8 kph (almost seven mph). That's considerably faster than the quickest human swimmer, and the penguins are also able to keep up these high speeds over far greater distances. To do so, they need to keep their plumage in very good condition, which is why, like other penguins, they spend so much time preening their feathers with their bill. They also lubricate them regularly, using oil produced by a special gland near the base of their tail, which again they apply by using their beak.

Those flippers – vestigial wings which are held down by the birds' sides as they walk – turn into powerful propellers once the bird is in the water. This, together with their powerful feet, enable them to swim very fast, and also to manoeuvre themselves when chasing prey or, in turn, being pursued by a predator. The power generated by the flippers is considerable, especially when you consider how small they are compared with the wings of an albatross, for example. One reason why the emperor penguin's flippers are so small in relation

'Like all penguins, the emperor is well adapted to a lifestyle in which they spend much of their time at sea.'

ABOVE

Gullies, produced by the accumulation of blowing snow between the sea ice and ice-shelf cliff, presented potentially fatal traps for adults and chicks alike.

to its body (they cover a lower surface area than those of the smaller king penguin) is that larger flippers would lose more of the bird's precious body heat.

Like all penguins, the emperor is well adapted to a lifestyle in which they spend much of their time at sea: in its case, in the freezing waters of the Antarctic Ocean. Their body is highly streamlined, with long and narrow flippers that enable it to dive to incredible depths – emperors have been recorded at depths of more than 500 metres (over 1750 feet) beneath the surface.

Emperor penguins also have other adaptations to their respiratory system, skeleton and metabolism, which together enable them to stay underwater for up to 32 minutes. Their body can still function with very low levels of oxygen, and they are able to slow down their metabolism so that they can stay down for so long. During a dive the emperor's heartrate may fall as low as 15 beats per minute, while it can also shut down any internal organs not required.

Finally, their bones are solid – unlike those of flying birds, which are hollow and filled with air, to enable them to get and stay airborne. Solid bones mean that the emperor can, when diving to such great depths, withstand pressure that may be 40 times greater than at the surface, which would kill a human being unless they were wearing specialist diving gear.

Courtship and pairing

By the end of April, three weeks or so after the first emperors began to arrive back at the colony, temperatures had dropped as low as minus 35 degrees, and as a result the sea ice was finally freezing solid. It was time for the next stage in the breeding cycle to begin: courtship and pairing.

In the vast majority of bird species around the world, the male takes the lead in courtship, using either sound or visual display to attract the female and then forge the pair bond between them. Despite being so different in many other areas of their lives, in this aspect emperor penguins are no exception to the general rule.

At this point in the breeding cycle, the male begins his special display. First, he stands stock-still and lowers his head to his chest. Then he takes a deep breath of the freezing air and begins to call: the sound lasting just a few seconds. He repeats the same action while walking slowly around the colony, until he manages at last to stir the interest of a female.

Like other birds with complex courtship displays, such as grebes, the next stage involves both the male and the female. They stand face to face, and then

During courtship,
the male and female
mirror one another's
movements and postures
to reinforce the pair bond
between them.

go through a series of coordinated movements, as if each is looking at the other in a mirror. When the male lifts his head up, so does the female; when he lowers it, so does she. Each pose is held for a minute or two, before they change position.

This goes on for some time, until the pair are fully bonded, after which they usually stay together as they walk around the colony. This is always a critical point in any pair of emperors' relationship: having committed to one another, they will be jointly responsible for raising a single, precious chick, a process that will take another six months or more. So, even some time after pairing up they will still engage in these bonding rituals, often uttering loud calls, to reinforce the link between them. The male and female will have to work closely together as a team, so being able to synchronise their actions is crucial.

WITH THE ICE getting thicker every day, it was now safe for the crew to walk and drive on; and, finally, Lindsay could get closer to the colony. Before the courtship began, he had got great footage of the birds coming to the colony, and moving around once they had arrived. But then the snow had made filming impossible, and the team had to wait nervously for the weather to improve.

Fortunately, soon afterwards, it did just that, with settled, fine conditions taking hold. Temperatures were still down to minus 35 degrees, though, so getting to the colony was tough – but finally the team found a ramp onto the sea ice and were able to drive their skidoos down and get close to the penguins for the first time.

They were almost too late but, fortunately, some birds had still not gone through the courtship and mating process, and Lindsay was able to get good footage of them doing so. "Emperors are such stunning birds, and their display is equally beautiful; slowly mimicking each other by raising and lowering their bill, and stretching tall to show off their golden neck patches to each other."

For Stefan, witnessing this intimate ritual was also a real privilege. "It was a very elegant, almost artistic, process. One of the most beautiful things to see was when they stood tall in front of each other, leaning their heads to one side and looking at each other. This really brought out the quite muscular nature of these birds, and also showed their immense size. They would stay in this pose for up to a minute, never moving an inch, until they slowly relaxed their body again and sank back to their normal size."

Will recalls that the birds' courtship was not as straightforward as you might expect. "One thing that struck me was how often the individual was rejected. The initial call and mirroring display would start, but quite often, after 30 seconds or so, one or other bird would simply walk off. The individual I was watching must have endured this rejection at least four times before I finally lost sight of her in the melee of penguins."

‘ Witnessing this intimate ritual
was a real privilege. ’

ABOVE

Mating between pairs is
often more comical than
romantic, as the male
tries to mount the female
on slippery ice.

OPPOSITE

Even in such a cold,
unforgiving place, tender
moments frequently
occur between males
and females.

The final stage in this part of the breeding cycle is mating. But compared to the ritualistic and aesthetically pleasing courtship ritual, the actual mating was something of an anti-climax; even at times comical to watch.

The team soon learned the signs that suggested mating was about to occur. First, the female would stand up straight, extend her flippers out to the side and then slowly descend onto her belly. Then she would lift up her tail, sending a clear signal to the male that she was ready to copulate. He would approach and begin to do so.

But this was not an easy task. "It was like watching a surfing lesson: the male would try to clamber onto her back, like a human being standing on a surfboard for the first time; but then he would lose his balance and fall off."

Once the male had finally mastered this precarious position, he would then begin to mate, lowering his body down so that his tail touched that of the female's. Then there would be a few brief rocking movements and the whole process was over, almost as quickly as it had begun.

IT WAS NOW the middle of May – equivalent to mid-November in the northern hemisphere – meaning that there were only five days left during which the sun would still peep above the horizon and illuminate the busy scene. On 21 May, the 'polar night' began here. For the next eight weeks, the sun would not rise at all. However, it would not be completely dark as, with the sun only just below the horizon, there were still a couple of hours of twilight each day, during which they could continue filming, using highly specialised light-sensitive cameras. But the lack of light – together with wind-chill bringing the temperature down to below minus 50 degrees – certainly did not make things any easier for the crew.

The eggs are laid

By the start of June, temperatures had dropped so low that the majority of the birds had huddled together for warmth at the centre, with just a few pairs around the periphery: either displaying to one another, resting on the ice or simply wandering around. There were also a few lone birds – presumably all unmated, and likely to remain so, having missed out at the start.

By now, the light had almost gone, and temperatures were down to minus 40 degrees on a regular basis. But it was in such cold, dark conditions that the most crucial stage of the breeding process would take place – a process that, if it went wrong, would mean that it would all be over until the following year.

First, it was time for the female to lay her egg. Emperor penguins are different from all other penguins – and most other birds – in that they do not build a nest. For a start, there are no materials with which to do so; and even if there were, an egg exposed to the outside air would freeze within minutes.

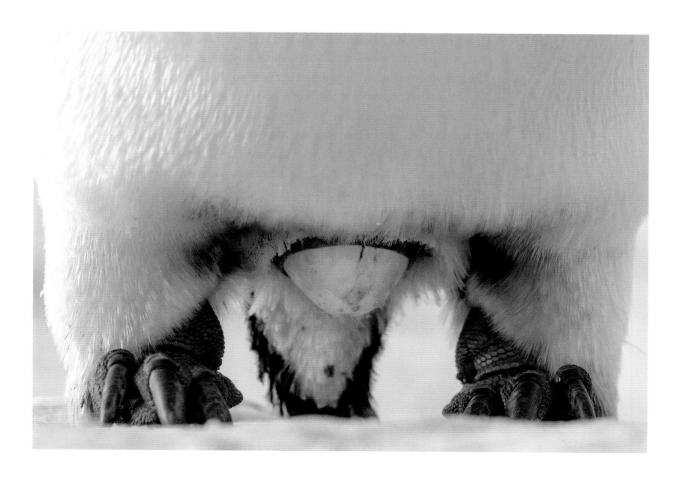

> ' Some biologists now consider emperor penguins to be a superorganism – in the same way as ant or termite colonies. '

The huddle

Inside the egg, the tiny chick was gradually growing and developing. It had everything it needed – at least for the time being – in terms of nutrition and because the egg remained permanently against the male's brood patch, where its surface was regularly in contact with the male's skin, it was kept warm.

For the male, however, things were not quite so comfortable. On (relatively) warm and windless days, when the birds can spread out, the situation was tolerable. But when bad weather brought high winds, these rapidly chilled the surface of the feathers and meant that the penguin lost heat far more rapidly. The crew could, at these times, retreat to the warmth and safety of their base, but the birds did not have that option.

So, they did the next best thing, and formed a huddle – in Stefan's words, "the emperor penguins' secret weapon against the cold". The birds stood in a rough circle, all facing inwards towards the middle. As new birds arrived from outside, they gradually worked their way in towards the centre, as the

As well as keeping the egg warm and safe, the males must maintain their own body warmth. They do so by huddling close together.

individuals there moved out so that the newcomer could take their place. The concentration of birds is truly extraordinary: they can squeeze in at densities of between 10 and 12 birds per square metre, leaving little or no room for manoeuvre.

As they do so, instead of losing their body heat to the freezing air, that heat stays within the huddle, so they stay warm. In the very centre, this can create a temperature of more than 30 degrees Celsius, vastly higher than the external air temperature of minus 40 or even minus 50.

This hugely reduces the amount of energy lost by the birds: typically, a male penguin, when exposed to the elements, will lose up to 300 grams (10.5 ounces) of their body weight each day; but when in the huddle this daily weight loss falls to just 120 grams (four ounces). That might make the difference between surviving the long Antarctic winter, or not. Huddling also explains why emperor penguin colonies are so large: if numbers fell too low there would not be enough birds to form an effective huddle.

Some biologists now consider emperor penguins to be a superorganism – in the same way that in ant or termite colonies, and in beehives, the

As each male moves just a few centimetres to readjust his position, the whole huddle reacts, as the movement ripples throughout nearly 4,000 birds. As a result, the mass of bodies is constantly on the move.

individuals act together for the collective good. One big difference is that in social insects many of the individuals are far more closely related to one another than are the emperor penguins, having all descended from the same queen. Another is that ultimately each penguin is still acting for its own good – in a sense, they simply have no choice: if they fail to cooperate, they and their unborn chick will die.

The huddle constantly rotates, as those in the centre move out towards the edge, because the birds can actually get *too* warm inside. So, after staying there for a while, and warming up, the emperors in the centre then have no choice but to move out towards the edge again, if they want to avoid overheating.

Stefan describes watching this process at work. "Eventually, temperatures will rise beyond what the penguins can bear, and the huddle will break open. It's a remarkable sight, because all of a sudden there is a lot of movement, the volume of the emperors' calls picks up, and a giant cloud of steam emerges from amidst the birds, while they start to spread out flapping their wings. You can really tell that they are nicely warmed up, feeling all comfy and relaxed."

Will was also surprised at just how much heat was generated from the birds. "I was taken aback by how much steam was produced when the huddle would

During winter storms, as the winds increase, those birds exposed to the full force of the weather will periodically make their way around the huddle edge, to the more sheltered side. As this process is repeated the colony is gradually pushed in the direction of the wind.

break open. On clear days, with the sun low in the sky, it looked like the lid being taken off a pan of boiling water. When there was no wind, the steam would rise upwards in a thick column, backlit by the sun – an amazing sight to witness in such bitter temperatures."

Lindsay often noticed a breakdown in this collective behaviour. "If just one bird wasn't happy, perhaps because it had become overheated, he could cause the whole group to break open, move, and then have to reform again. The way birds leaned and pushed to pack as tight as they could made it look like a massive rugby scrum. Some even took run-ups before trying to squeeze into the smallest gap!"

At times, especially on windy days, they observed a variation in the usual strategy. The penguins standing on the windy side would walk around the huddle onto the sheltered side, exposing other birds who would then follow suit. By this process, the whole huddle would move away from the wind, with every bird getting to spend some time on the sheltered side, until they had to walk around the huddle again.

Eating for two

ABOVE

As the females enter the water, muscles in the skin contract, producing streams of bubbles; air, previously trapped between the feathers for insulation, is squeezed out so that the birds are more streamlined.

OPPOSITE

The females must fatten themselves up as they prepare to return to feed their newly hatched chick, diving to depths of nearly 500 metres (1,600 feet).

The female emperor penguin, having laid her egg and then deposited it with the male, urgently needed to get back to the sea, where she would be able to find food. This was a very long walk which she had to undertake when she was at her lightest and weakest, having put so much of her energy into laying that egg.

At last, however, she managed to reach the edge of the ocean, where, along with the other females, she dived into the water. Emperor penguins feed mainly on fish, squid and krill – the small crustaceans that form such an important part of the diet of many Antarctic mammals and birds. However, the Victorian explorer James Clark Ross (after whom the Ross Sea is named), examined emperor penguins whose crops contained up to 450 grams (one lb) of small stones and pebbles. At the time, it was thought that these might act as ballast when the birds dive for food; but now we know that they are used to aid their digestion.

Typically, emperor penguins dive to depths of up to 50 metres (160 feet), pursuing their prey using those manoeuvrable flippers to follow their twisting and turning movements. Sometimes they swim beneath the surface of the ice floes, where some of the prey is hiding; at other times they will head deeper.

At first, like the others, this female needed to catch as much prey as she could, in order to return to her original body weight. But she was not just feeding for herself. Having eaten her fill, she then needed to hunt some more, to fill her stomach with as much food as possible – this time to bring back for her chick.

Stefan recalls what happened next. "Upon our descent onto the sea ice and getting closer to the colony we could finally hear the call of the chicks." All over the colony, each chick was breaking through the thick shell of its egg, using a special 'egg-tooth' – a sharp protuberance on the unborn chick's bill.

Once the chick had emerged, the female pointed her bill downwards towards the ground. The chick instinctively responded by pointing its own bill upwards, while making high-pitched begging calls, which stimulated the female to regurgitate food from her stomach. In cases where the female does not return before the chick is born, the male is able to produce a secretion from his own stomach that will keep the youngster alive for a few days. But if the female has been killed by a predator, or died on her journey to or from the sea, from starvation or exhaustion, then the chick is doomed.

Witnessing this ritual was a moment to reflect. "It's hard to believe that these small and fragile creatures will grow into majestic and stoic penguins, facing the harshest living conditions on the planet on a daily basis. Nature just never ceases to amaze."

Once the chicks are nearly two weeks old and covered in a warm down, they spend part of their day exposed on their mother's feet. From this safe perch, it isn't long before they start getting to know one another.

OPPOSITE

Weighing just 400 grams (less than one lb) when hatched, the chick's weight almost doubles every two weeks, thanks to the female's ample supply of food.

For Lindsay, having spent so much time filming the males in almost total silence, the change in the atmosphere of the colony was especially noticeable. "The colony went from a silent huddle of males trying to keep warm, to a deafening mass of males, chicks, and females calling to find their mates."

When the chicks were born, each had a downy, grey plumage, with black and white markings on their face, as Stefan recalls. "The chicks were incredibly adorable, especially after just having hatched. They were still mostly naked and only covered by a very thin layer of down which still exposed a lot of skin to the cold temperatures. Their little pear-shaped bodies were tucked away under the brood pouch of their fathers, with their tiny feet standing on top of those massive feet." Lindsay, too, noticed the contrast, recalling that in comparison with the tiny feet of the chick, the adult's feet looked like they belonged to enormous prehistoric beasts.

Disaster strikes the colony

ABOVE

The unforgiving landscape constantly shifts, creating new hazards. A chick stuck in a steep-sided gully is unable to find a way out, with the rest of the colony just metres away.

Just three weeks after the team had witnessed the first chicks hatching, the notoriously unpredictable Antarctic spring weather sprung a surprise, as a brutal storm hit the colony. As the chicks were still so small, they feared the worst.

After five days the weather finally began to take a turn for the better, so the team headed out from the warmth of their base to survey the damage. As soon as they arrived, as Will recalls, they could see just how disruptive the storm had been. "We could see a lot of chicks wandering aimlessly around on their own, obviously separated from their parents, some so small still that if they didn't find help we were quite sure they would not survive. Small mounds of snow marked chicks that had already succumbed, and the more we looked, the more small, snowy bumps we saw."

A large proportion of the emperors and their chicks were now on the ice shelf above the sea ice, as the wind had pushed them up the ramp of snow that links the two. There were deep gullies all the way along the boundary between the ice shelf and sea ice, into which any unwary penguin could easily fall.

When the team arrived at the edge of the ice cliff, they were shocked by what they saw. In Lindsay's words, "It was carnage. As we looked down the terrifyingly deep cracks in the ice we saw adults, still with chicks on their feet, peering back up at us, unable to escape. It was heartbreaking."

The problem facing the adults was that with the chicks balanced on their feet, and protected by the brood pouch, they were unable to use their claws to climb up the steep sides of the gully as, if they did, they would have to abandon their precious offspring. Meanwhile, birds without chicks were moving in and out of the gullies, seemingly intrigued by the calls of the trapped adults and chicks within. And some chicks had already lost the battle to survive, and were strewn around the gully floor.

Time and again, the trapped adults with chicks tried to scramble up the sides, and time and again they failed to do so. As Will recalls, "It looked hopeless. We spent all day willing the adults to find a way to make it out, but none did. Then, finally, just as the light was fading one female made a bold move on a new route, steadily inching up the sloping walls, using her beak like an ice axe, until she finally managed to drag herself over the lip, amazingly still with her precious chick safely on her feet!"

BELOW

After a heroic effort, a mother manages to escape with her chick.

Growth and independence

BELOW

By December, the chicks are almost half the weight of their parents. Over the coming weeks they will receive their last feeds and begin to moult in preparation for several years at sea.

After the female returned to take over care of the newborn chick, it was the male's turn to head back to sea. He had not fed at all for almost four months, and as a result had lost a considerable amount of weight. He also needed to waddle or skid across the ice for days on end; and when he finally reached the edge of the ocean, he also had to avoid predators, catch enough prey to restore his depleted bodily resources, and then – just like the female – eat more, to bring back food for his hungry chick.

Once he had returned, the female handed back the chick and headed off again. During the next seven weeks or so, the male and female took turns to return to the ocean to catch food to bring back for the growing chick, swapping over each time they did so.

When the chick was still young, it spent most of the time inside the brood pouch. Sometimes it would stick its head out, taking a look at what was going on in the colony without having to endure the freezing temperatures. Lindsay recalls that the females would walk around the colony, as if showing off their chick to the neighbours.

At nearly two months old,
the well-grown chicks
no longer rely on their
parents for warmth.

As the chick grew bigger and bigger, it soon became too large to stay there, and the adult finally pushed it off its feet and onto the ice. By the time it was six weeks old, by early September, it was fairly independent, and able to wander around the colony.

In another example of the emperor penguin's unusually cooperative behaviour, both parents then left the chick in a crèche, together with the other youngsters, where they too would huddle together against bad weather and storms. By now, the sea ice had begun to melt, which meant that the adults' journey back to the ocean was a little shorter than before. And with both parents going off to feed, they were able to bring back enough food for their rapidly growing offspring.

Will recalls that at this stage the scene resembled parents picking up their children in a school playground. "The colony itself was spread out over a square kilometre or more of the sea ice, forming several smaller groups. The returning adults would move from one group to the next in an attempt to find their chick, heading to the small crèches of chicks and then standing on the edge, calling and waiting to hear if their own chick was amongst them."

But not every pair of penguins had been so fortunate. Some eggs failed to hatch or the chick died soon after birth, following which the pair split up and male and female went their separate ways. Yet the instinct to raise a chick

Even though the sacrifice and endurance for all the adults is the same, not all are successful. But driven by hormones, females do not give up their parental ambition, looking for any opportunity to steal the offspring of a neighbour.

continues to be very strong, especially in the case of the female, so sometimes a female who has lost her egg or chick 'kidnaps' another. But if they no longer have a mate, they are unable to leave the chick to find food, so the chick may not survive.

On one occasion, the team observed and filmed an extraordinary scene. A recently reunited couple, with a healthy chick, began the process of passing the chick from the male to the female, so that she could feed it with regurgitated fish and squid. But just as they began to do so, a group of unpaired birds on the periphery of the colony descended on them, and tried to grab the helpless chick.

"By now, the chick was scrambling to find safety with the other adults' beaks coming down like spears all around it, which was really harrowing," Will recalled. "No more than a few days old, this chick was adrift from its parents' care, and fighting for its life." A few moments later, this scrum of birds separated, to reveal the parents of the chick facing one another, pouches empty, and their chick nowhere to be seen. "A mere few feet away, the kidnapper unfurled its pouch to reveal a clearly rattled chick, its newly adopted offspring, covered the chick again and shuffled off into the crowd. The couple's breeding for this year had been ended without warning by one of their own, revealing a very different, almost sinister, side to colony life."

BUT MANY OF the chicks had survived. By December, the pack ice breaks up enough for the sea to be very close to the breeding colony. With open water nearby, it was time for the chicks to leave their parents and at last become independent. They were now beginning to moult, losing their fluffy grey down. This moulting period can last over a month, during which the birds are unable to feed. But once they have their smart new feathers, they are then able to go to sea and hunt for food, just like their parents.

It is now a time of plenty, and for the emperor penguins, another long and arduous breeding season is finally over. Yet within less than four months, they will have to start the whole process all over again. And five years later, those tiny chicks that managed to survive their early days on the sea ice will also return – this time to breed, and raise a family of their own.

CHAPTER FIVE

Tigers

The sun sets over Bandhavgarh National Park, home to one of the highest concentrations of tigers in India.

AS DAWN BREAKS, a tiger heads out from her den into the forest of Bandhavgarh National Park in Madhya Pradesh, central India. She moves slowly, stealthily, and always very quietly. This tiger is, as always, on the lookout for a victim: a sambar or chital deer, perhaps, or a wild boar. Tigers are not particularly fussy eaters, and will hunt a wide range of prey, though they usually prefer a larger animal, as this will obviously provide more food.

She is a lone hunter, acutely sensitive to any signs that prey might be within reach. Despite the high density of grazing animals here, she must be patient. Prey animals have co-evolved with tigers to be constantly on the lookout for danger, and it only takes a single misstep – cracking a twig underfoot or inadvertently revealing her silhouette in a gap between trees – to give away her presence, and allow an observant creature to sound the alarm and get away.

But today, her chances are looking good. A herd of chital deer – a dozen animals in all – is browsing the lush green foliage in a forest clearing. She has vision up to six times as acute as that of a human being, enabling her to focus clearly on the shadowy figures of her potential victims as they continue to feed in the early morning light, unaware of the danger so close by.

Tigers often prey on sambar deer; the deer appears to have spotted the predator's approach, so this time the hunt is unsuccessful.

The closer she gets, the more likely that one of the deer will sense her presence and sound a warning. Indeed, the majority of her hunting attempts do end in failure: if a tiger realises it has been seen, it will usually give up the chase before it has even begun.

Her luck appears to be holding. She inches closer and closer, moving so slowly she hardly seems to be making any progress, yet gradually getting nearer to her target, a big male on the edge of the group. She must be very careful – one lunge from his antlers could prove lethal. But the potential reward of such a meal still outweighs the risk that she might be injured in the attack.

Finally, after half an hour's patient pursuit, she is within a few metres of her victim. She crouches down and leaps forward, knocking the chital off balance, grabbing the animal by the throat with her powerful jaws and then holding it down with her front legs. The deer was doomed even before it hit the ground; a few moments later, its air supply cut off, it dies.

THIS TIGER IS Raj Bhera, a mature, fully grown female. At the start of our story she is five years old, strong and experienced; the ruler of a tiger dynasty that goes back for at least a century, perhaps longer. She has a grown-up daughter named Solo, whose territory is close to hers.

Dhruv Singh, who comes from this area, knows that these tigers have a long history. "Raj Bhera lives in the shadow of Bandhavgarh fort, an ancient city once home to maharajahs. Her dynastic rule echoes that of the people who once

OPPOSITE

Raj Bhera may be fully mature and experienced, but she still faces enormous challenges every day of her life.

BELOW

During the day, tigers often rest up out of sight; here, Raj Bhera sits in a bamboo glade.

lived here. We've followed her ancestors ever since tourists started coming to the park, and now her story is one set in a changing India that is straining the relationship between human and animals as never before."

But although Raj Bhera is at the height of her powers, like any wild creature she faces constant challenges. She has a litter of cubs hidden away in a den, in a remote and hilly part of the national park. Her task is to raise these cubs from birth, all the way through to adulthood and independence. She will only achieve that goal if she is able to find enough food for her and them, while at the same time keeping herself fit and out of danger, and defending her territory against intruders. And she will have to do this all on her own, because once they have mated with the female, male tigers play no part in raising their family.

This is a story that reveals the tiger's power, as the top predator on her home patch; her tenderness as the mother of a litter of cubs; and, despite her size and power, her vulnerability. We also discover that unless we are willing to make room for tigers in one the most populous nations on the planet, where they must find space to live amongst more than one billion people, their future looks very precarious indeed.

Tigers and their lives

In appearance, a tiger is quite simply unmistakeable. No other big cat has that distinctive striped coat. Although the stripes may seem obvious when the tiger is out in the open, as soon as it walks through its jungle habitat, they help it blend seamlessly into the dappled sun and shadow, and make it very hard to see.

Of all the big cats, only the lion rivals the tiger in terms of size. A big male tiger can reach a total length of four metres (over 13 feet) from its nose to the tip of its tail, though most are smaller, at around 2.2–2.5 metres (7–8 feet) long, with a shoulder height of about one metre (3.3 feet). The tail makes up around one-third of a tiger's total length.

A male Bengal tiger may tip the scales at over 300 kg (660 lbs), the equivalent of well over three adult male humans of average weight. However, they are generally lighter than this, mostly ranging from 200 to 260 kg (440–

The famous striped coat of the tiger evolved for camouflage; once the animal is in a dappled forest environment its benefits become clear.

570 lbs). While females are usually about one-third smaller than the males, and weigh between 100 and 160 kg (220–350 lbs), they are still sizeable animals. Both males and females are built for hunting, with muscular shoulders and forelegs, a thick neck, sharp and retractable claws, and a powerful jaw full of huge, razor-sharp teeth.

Principal wildlife cameraman John Brown, who has spent hundreds of days in the field observing and filming tigers, was struck by their ability to change shape, depending on which angle you observe them from. "I was always amazed by how the tigers seemed almost two-dimensional. Viewed from the side, they have a real mass and muscularity, but they are incredibly slim in the hips and shoulders, so they almost vanish when they are walking towards or away from you."

The colours and patterns of a tiger's coat vary considerably between individuals, allowing scientists and filmmakers to identify particular animals.

However, John found that they could not always tell the tigers apart. "Even though a tiger's stripe pattern is as unique as a fingerprint, and we kept

Tigers are excellent swimmers, and will also often rest in water to cool down during the heat of the day.

reference images in the filming vehicles, it still proved surprisingly difficult to positively identify individuals. During the filming period, the territories of Raj Bhera and the surrounding females were very fluid, and we had to be careful not to jump to conclusions about which individual was the one we were looking at, simply based on where we saw them."

We usually associate tigers with jungles, and in India that is indeed their preferred habitat. However, they are very adaptable creatures, able to live and hunt in a wide range of different biomes including river basins, reed beds, scrub, montane forests, and on the edge of mountains, though always with some tree cover. They are also able to cope with temperatures ranging from a low of minus 34 degrees Celsius in the Siberian winter, to well over 45 degrees in the baking Indian summer. And they can live anywhere from sea level to altitudes of up to 4,000 metres (13,000 feet) in the Himalayas.

Unlike lions, which are social animals based in a pride, tigers live solitary lives for most of the year, apart from a brief period when males and females get together to mate and, of course, when the female is raising her litter of cubs, as Raj Bhera is now doing. However, they do interact with one another frequently, especially in an area such as Bandhavgarh, where territories are cheek-by-jowl. The tigers being filmed were always aware if any other tigers were in the vicinity, and would respond accordingly, either by avoiding intruders or, sometimes, confronting them.

Hunting and killing

Daytime temperatures may have risen to over 40 degrees during the height of summer, yet Raj Bhera is still able to drag a large male sambar deer back to her cubs.

Along with other big cats, tigers are apex predators; unlike, however, the African trio of lion, leopard and cheetah, an adult female tiger like Raj Bhera has no real rivals. Apart from the much smaller leopard, which usually hunts smaller prey, all the other major predators have been wiped out from the vast majority of the tiger's home range across Asia.

Like other tigers, Raj Bhera is a mostly nocturnal hunter, covering anything between 7 and 30 km (5–20 miles) a night in her travels. She uses the classic technique of 'stalk-and-ambush', in which she creeps slowly towards her victim and then pounces at the last possible moment.

Even so, most hunts are not successful. But whether they are or not, they are always a thrill to watch. "As we watched Raj Bhera begin to stalk her prey, it really felt like there is a build-up of static charge in the air," recalls John Brown. "We could certainly feel the tension building as she moved into position. There was always a huge sense of anticipation among those of us in the camera car, and it always struck me as incredible that the target of the stalk could be oblivious to this."

Yet most of her victims are totally unaware of her presence until it is too late, and she has entered the final stage of the hunt: the chase. Along with other ambush predators such as lions, tigers can accelerate very rapidly to reach a top speed of up to 65 kph (over 40 mph) in a few seconds. But they tire very quickly, so cannot usually pursue their prey over any distance longer than a few dozen metres.

Yet there are occasional exceptions to this rule. John Brown recalls that, on one occasion, the team saw Raj Bhera run close to 200 metres at full pace, and

'A large adult tiger like Raj Bhera can eat up to one-fifth of its bodyweight from a single kill.'

successfully catch and kill a sambar. "Was this a sign of desperation, given that she urgently needed to make a kill, or was it just an indication of what a skilful hunter she is? She had watched two sambars for nearly an hour before the charge, so maybe she had some sense that the individual she singled out was vulnerable?"

Like all opportunistic predators, Raj Bhera will hunt a wide variety of prey, taking anything from the largest deer (such as a big male sambar, weighing at least twice as much as they do) to smaller mammals such as muntjac deer, hares and porcupines. But in the forests of central India, the vast majority of prey animals are large- to medium-sized deer, which she prefers as, if she succeeds, she gets more meat for her efforts. These are often hunted at waterholes where they come to drink, as their regular and predictable appearance means that she can lie in wait until they arrive, then attack when their guard is lowered. For Raj Bhera, the dry season always brings a concentration of prey at the few areas of standing water still remaining.

The method of killing varies, depending on the size and strength of the victim. When hunting small animals, she will simply grab them and bite their upper neck with those fearsome canines, which breaks the spinal cord immediately. She'll then consume their flesh in one short sitting. But larger animals, especially male deer or wild boars with potentially dangerous antlers or tusks, need to be approached much more carefully. Having brought down her target, she will bite the animal on the throat and then use her powerful forelegs to force it down to the ground and hold it still. She will then suffocate the prey.

The ability to hunt animals far bigger and heavier than itself makes the tiger unique amongst large predators – lions and cheetahs do go after large prey, but usually they hunt as a team rather than on their own. Not so the tiger, which always hunts alone. Once a tiger has successfully killed, it will usually drag the body of its victim into cover, so it can feed without being disturbed and stop any of the meat being taken by other predators – including rival tigers.

When feeding, Raj Bhera will gorge herself on the meat, starting at the fleshy hindquarters of the animal, and using her side teeth to rip off huge chunks of flesh, before then moving on to the rest of the carcass. A large adult tiger like Raj Bhera can eat up to one-fifth of its bodyweight from a single kill, a process that may take two or three days.

Having finished feeding, she will often cover up the remains of the carcass, using whatever material comes to hand including leaves and grass, in case she needs to return later to feed some more. After that, she will not normally need to feed again for up to a week, so will retire to rest and digest her meal. But now that she has cubs to feed, she will need to kill again soon.

Bandhavgarh and its tigers

Bandhavgarh National Park is set amongst the forested Vindhya Hills in the state of Madhya Pradesh, in eastern central India, roughly 800 km (500 miles) southeast of the country's capital Delhi. It is home to one of the largest – and densest – populations of tigers anywhere in the world.

Like other national parks in India, Bandhavgarh has a very varied terrain and vegetation with jungle, forest edge and grasslands, and reaches a height of 800 metres (2,640 feet) above sea level. The park is home to healthy populations of a wide range of other wild animals, including large populations of deer, as well as sloth bears, leopards, Indian elephants and several hundred species of birds.

Along with so many areas in India where wildlife still manages to thrive, Bandhavgarh was once a place where the ruling maharajahs and their guests hunted tigers. Ironically, it was only because the area was set apart for shooting that the tigers and their habitat managed to survive.

Fortunately, during the course of the twentieth century, people became more aware of the growing plight of India's tigers, and eventually hunting was stopped. The whole area was handed over to the state government, and opened to tourists. Today, thousands of visitors come here from around India

The view of Raj
Bhera's territory from
Bandhavgarh Fort.

and beyond, travelling around the park in vehicles in the hope of catching at
least a glimpse of the park's tigers – and in doing so they contribute much-
needed revenue to the local economy.

The National Park itself was created in 1968, and originally covered a total
area of 105 square km (40 square miles). In the early 1980s, this was more than
quadrupled; since then, the park has expanded further, and a buffer zone has
been added, taking the total area where tigers are now protected to more than
1,530 square km (600 square miles).

Even so, the high density of tigers here means that each animal's territory
is much smaller than would be ideal, often leading to conflicts between a tiger
with an established territory and another – often related to the incumbent –
that is trying to create a territory of its own.

Bandhavgarh was the ideal location to film tigers, for several reasons. First,
the high number of tigers increased the likelihood of being able to track and
film these elusive animals. Second, the long history of wildlife tourism in the
area meant that the tigers were habituated to the presence of people. Finally,
a network of tracks around the park meant the team could follow the tigers –
either in vehicles or, for filming, on the backs of elephants.

But as director Theo Webb points out, none of this would have been
enough unless the team had an animal they could follow as it went about its

Cameraman John Brown
takes advantage of the
view to film the setting
sun from the fort.

daily life. "Most importantly of all, we needed to find an experienced female tiger, in a reasonably accessible location, and which had a strong chance of getting pregnant and having cubs during our three-year filming period."

The only tiger that was a realistic candidate for filming was Raj Bhera. But the task was still a very tricky one, as the team found out the hard way on their very first filming trip.

At that stage, the team had heard reports that Raj Bhera had definitely mated, and so there was a good chance that she would soon give birth to a litter of cubs. But their best-laid plans were scuppered when they discovered that this time she was not, in fact, pregnant.

By the time of the second shoot, just a few months later, there was much better news. Not only had Raj Bhera become pregnant, but she had also given birth to cubs. The bad news was that – as often happens with female tigers once they have a family – the star of the show had become very elusive. To keep her new family as safe as possible, she had retreated to one of the most inaccessible corners of the park, keeping out of sight for much of the time.

Fortunately, the park is patrolled by Forest Department guards on elephant back who constantly monitor the tigers to prevent poaching. Neelam Singh, the guard and mahout in control of the elephant, was able to help Theo in his search. This was the only practical means of accessing the hilly, off-road areas where the den was likely to be. It is also by far the best way to find tigers, which are used to the presence of these mighty animals, and will allow a much closer approach than if the trackers used a motor vehicle.

BELOW

Mahout Neelam Singh used his skills and expertise to guide cameraman John Brown; this was after they had successfully filmed a tiger kill.

Two weeks into the shoot, Theo and Neelam had more or less given up hope of ever finding Raj Bhera's hidden den, as Theo recalls. "To find where she had chosen to have her cubs, we first needed to find Raj Bhera, and then to follow her through really difficult terrain, hoping that she would finally lead us to her den. We kept sighting her, but almost as soon as we did, she would melt into the forest and we'd lose her. It was incredibly frustrating."

Finally, though, their luck changed. Early one morning, the sharp-eyed Neelam found some fresh tiger tracks going up a steep hillside. Soon afterwards, Theo spotted a small cave beneath a rocky overhang. It looked promising.

Neelam stood up on the back of the elephant and, a few metres away, he could see Raj Bhera, lying on the ground at the mouth of the cave. Moments later, the cubs appeared, aged just two or three weeks old. Theo and the crew were delighted that their hunt was finally over. "Neelam and I couldn't believe that we had found it – this really was like looking for the proverbial needle in the haystack. We realised, as we watched the cubs, that we were the first human beings ever to see them – it was a really emotional and incredibly exhilarating moment."

Eventually the team discovered that there were four cubs in all. Raj Bhera had found the ideal den site to raise her family: remote, inaccessible, and out of the way of any danger. This was the start of the most crucial period of her life: raising her tiny cubs all the way to maturity.

Tiger breeding

Munga, the father to
Raj Bhera's cubs.

OPPOSITE

Tiger mating can be a
noisy and sometimes
violent affair, as the male
often grabs the female by
the scruff of her neck.

Raj Bhera, like other female Bengal tigers, had first attained sexual maturity
some time between three and four years old. At that point, she had come into
heat, sending a clear signal to any male tigers in the vicinity. Later, she had her
first litter of cubs.

The larger, older males (which reach sexual maturity roughly a year later
than the females) roam over a wide area looking for a mate. When one finds a
female in heat he must hurry, as she will only be receptive to his advances for
between three and six days. Meanwhile, he must keep a sharp lookout for rivals –
a female tiger's scent carries far and wide, and may attract more than one male.

As with other big cats, tiger mating is a rough and often noisy affair: the
male will grab the female by the scruff of the neck at the point of climax, which
can make the encounter appear quite violent. But there is a good reason for the
male's need to dominate: he must ensure that the female does not bite him,
or engage him in a fight in which one or the other might be left badly injured.
Nevertheless, as soon as copulation is over the female rejects the male.

Like lions, tigers copulate frequently to ensure that the female's egg is
fertilised: they may mate as many as 50 times in a 24-hour period. Even so,

Raj Bhera at the den
with her young cubs.
Female tigers show
great gentleness and
intimacy towards
their offspring.

there is a lower-than-even probability that the female will become pregnant – in some cases, as small as a one-in-five chance.

Mating is most frequent in the winter months, but can occur at any time of year. After it is over, the male and female part ways, and he will probably never see the offspring of this brief but crucial relationship.

ONCE SHE HAD got pregnant for the second time, Raj Bhera had just three months before she would give birth to her litter of cubs. Towards the end of this period, she began to seek out a place for her den, which would need to be hidden away in a cave or amongst rocks, where she would be able to keep her cubs safe.

Here she gave birth to her litter of four cubs – three males and one female – which were born blind and helpless, each weighing just 700–1400 grams (1.5–3 lbs). Raj Bhera's litter was fairly standard for tigers. This was her second litter: she had previously also given birth to four cubs, but that time there were three females and one male.

Her cubs, whose eyes open at roughly a week to a fortnight after they are born, will stay in the safety of the den for eight weeks or so, being fed on their mother's milk. They will grow very rapidly indeed: by the time they are four weeks old their weight will have increased fourfold. After a couple of months, Raj Bhera will gradually start to wean the cubs onto solid food, though she will continue to produce milk until they are about six months old.

The choice of a safe and secure den site is crucial for the future of these tiny creatures: any passing male tiger that is not the cubs' father will not hesitate to kill any youngsters he comes across; that way he prevents a rival male from succeeding. A female who has lost her cubs will generally come into season again soon afterwards, to maximise her chances of mating and raising a new family.

RAJ BHERA'S DEN site was out of the way of any males and with a good vantage point for her to protect her four young cubs. But once they reached the age when they were able to venture outside the den, she had her work cut out to keep them safe and fed.

Tiger cubs are keen to explore their surroundings – indeed, like all young mammals, this is the way they learn – so she had to keep a constant eye out for danger. Cubs also enjoy play-fighting, which again is a crucial stage in teaching them the skills they will soon need to make a life on their own, and Raj Bhera's were no exception.

Sadly, the chances of any one cub reaching maturity are less than fifty-fifty. Attacks by males are not the only cause of death for young tigers: they may die from an accident, bad weather, in conflicts with other predators such as leopards, by being killed by humans, and especially from starvation.

'The cubs, whose eyes open at roughly a week to a fortnight after they are born, will stay in the safety of the den for eight weeks or so.'

Rivalry with Raj Bhera's daughter

With a new family to look after, Raj Bhera couldn't patrol her territory as frequently or as diligently as she had previously done. As a result, her good fortune started to run out: a rival female pushed into a key part of Raj Bhera's territory. This intruder was no stranger, but her own adult daughter, Solo.

Now nearly three years old, Solo had moved into one of Raj Bhera's best hunting areas, which supported a very high density of grazing animals, including large deer. There, Solo was able to take her mother's prey, which potentially put Raj Bhera's new cubs at risk, as they needed all the food Raj Bhera could get.

John Brown remembers this being a real touch-and-go moment for Raj Bhera. "We could really sense her dilemma: should she relinquish resources

Female tigers usually avoid confrontation, but when their territories are so close they inevitably encounter one another from time to time.

and move away, in order to protect her cubs, or stand her ground against her own daughter and risk the consequences of a serious injury? We were all so enraptured by her at this point that it was not pleasant to watch the story unfold, as it didn't seem to be heading towards a happy ending."

Fortunately, the cubs were growing fast, which made Raj Bhera's life marginally easier. When they reached the age of three months, they were at last old enough to join her daily excursions to find food, enabling her to venture over a wider area to hunt. They would follow her at a safe distance, watching closely as she tried to track down any potential prey; and staying as quiet as they could when she finally did so. Cubs learn everything from their mother, so when she managed to make a kill they would watch carefully, before joining in the feast.

The more Raj Bhera and her family wandered away from the den and explored the further reaches of her territory, the greater was the chance that they would encounter Solo. And that's exactly what happened. Normally, adult female tigers will do their best to avoid one another at all costs, as

ABOVE

When Solo did challenge her mother, Raj Bhera was soon able to force her into submission.

fights can prove fatal. But this time there was no option, as Raj Bhera and Solo eventually came face to face. It happened when Raj Bhera had made a kill and, unbeknownst to her, Solo was watching. Gradually, Solo edged nearer and nearer, until finally her mother could no longer tolerate her presence. Now that Solo was fully grown, there was a possibility that she might take the kill from Raj Bhera.

The more experienced Raj Bhera knew exactly what to do: she had to show Solo who was the boss. And yet, initially, she momentarily hesitated, realising perhaps for the first time that her daughter had grown into a real rival. At this stage, it really did look as if the confrontation would result in a full-scale fight between mother and daughter.

Raj Bhera had to take action, and following that brief hiatus she raised herself up to her full height to confront her daughter. Solo immediately realised that she was facing an older and stronger animal, and adopted a submissive pose, rolling over onto her back. For now, at least, Raj Bhera had won. Her age and strength meant she was still in charge.

But there would soon come a time that Solo would start her own family. With hungry mouths to feed, she would then have no choice but to enter her mother's territory once again. By this stage in her life, she would be far more experienced – and so next time she might win the battle for supremacy between one generation and the next.

Tracking the Tigers

BELOW

To get moving shots alongside the tiger family, as they walked through the dense jungle, the crew used a specially designed gyroscopic mount.

Tigers may be one of the biggest animals here in Bandhavgarh, but that did not make them easy to film. Unlike other predators, such as lions and wild dogs, which give chase out on the open plains, tigers live and hunt in dense forest, so it's very hard to track and follow their movements.

So the team recruited expert help, in the shape of experienced guide and tracker Digpal Karmawas. As with all wildlife filming, the skills, expertise and local knowledge of people like Digpal and his colleagues was an essential factor in whether the team would succeed in filming the tigers at all.

At dawn each day, Digpal would take the team into the park, not knowing at this stage where Raj Bhera and her family would be. They would then split up,

Naturalist and spotter
Dhruv Singh compares
two images of tigers, to
make sure that the film
crew have identified the
correct animal.

One of the park's
elephants on anti-
poaching patrol.

to look for any signs of the tigers' recent presence. This is as much an art as a science, as Digpal points out: "Tigers are creatures of habit, but that does not make them easy to find. They have favourite paths and waterholes that they use but they always retain the ability to surprise, appearing where and when you least expect them to. Anything is possible in the jungle!"

Being ambush predators, tigers habitually hide in or behind trees or scrubby vegetation, and often the only evidence of a kill is the sound of the dying victim as it expires. The best way to track them was to use a three-stage approach.

The first step was to find their fresh tracks, which indicate that a tiger has been patrolling this part of their territory in the past few hours. Then, Digpal would listen for alarm calls from deer and monkeys, which would help pinpoint the tiger's current whereabouts. Even then, they are easy to miss: the third and final part of this process is the need for sharp eyes to spot the tiger in thick undergrowth.

But even when the team did manage to find a tiger, it was often almost impossible to see, let alone film. "They were so well camouflaged, especially in this dense undergrowth", explains Theo. "Their stripes blended in exactly with the bamboo, so it was often very hard to spot them."

The park guards on elephant-back could find the tigers even in the long grass.

For John Brown, the elusiveness of tigers was a real eye-opener. "In over 20 years of wildlife filmmaking, I'd never spent more time looking for the subject, and so little time with it visible through the viewfinder. We'd regularly go for days without seeing Raj Bhera, and even when we found her, the majority of the time the situation was unfilmable."

Using elephants did help the team get close to tigers, and also to follow them off the forest tracks, where a motorised vehicle could not go. Yet even when they came across Raj Bhera, in a position where they might be able to get some footage of her, their problems were not over.

The swaying movement of the elephants meant that it was impossible to film directly from elephant-back, as the resulting footage would be unusable. However, the strict park rules meant that no one was allowed to dismount from the elephant and film from the ground; this would be far too dangerous in such a densely forested habitat, where a tiger could easily be lurking within pouncing distance.

And although the den site Raj Bhera had chosen was good for her, its distance from the camp and the difficult terrain around the den made it very problematic for the film crew. One ingenious solution was a 20-foot long tripod, which they could place on the ground from elephant-back, and then retreat, so that John could then operate the camera from a distance, using a remote control.

Hanuman langurs are always curious animals. Here they are looking at a remote camera installed by the crew.

Tigers are naturally curious, always ready to investigate anything unusual they come across. So when the team put out camera traps in strategic places around Raj Bhera's territory, so they could capture her as she patrolled this large area, they expected her to come and take a look. What they did not bargain for was that the cubs would take such an interest that they would push one of the cameras over so that it fell into the water.

But despite these local difficulties, the team did finally manage to get some wonderful material of Raj Bhera and her cubs bathing in a nearby pond, playing together and at the same time learning important lessons about their new world. They were seen chasing cormorants and other waterbirds, and seemed to be enjoying playing in the water just like kids in a swimming pool.

For Theo, this made up for all the previous disappointments. "Camera traps are always very hit-and-miss – usually miss! But sometimes they are the only way to obtain this kind of intimate, close-up view of natural behaviour. And they gave us a real insight into the secret, hidden world of our tiger family."

‘ By the time winter began to set in,
Raj Bhera's four cubs were becoming
more and more independent. ’

Winter encounter

Dawn breaks across
Bandhavgarh National
Park, home to Raj
Bhera's dynasty.

One of Raj Bhera's three
male cubs. Not yet fully
grown, he still relies on
his mother for food.

Being hundreds of miles from the coast, and reaching an altitude of 800 metres
(2,640 feet) above sea level, Bandhavgarh has a continental climate, marked by
extremes of heat in summer and cold in winter. Temperatures can easily reach
45 degrees Celsius in midsummer, but on winter nights the mercury falls close
to zero, producing occasional frosts; and though it rapidly rises during the day
to 20 or even 30 degrees, the nights remain cold for the whole season.

By the time winter began to set in, Raj Bhera's four cubs were nine months
old. They were becoming more and more independent, with the family dynamics
beginning to change. At this time in their lives, when their mother caught her
prey, she no longer stood aside to let them feed first – instead, it was every
individual for themselves. That way they would soon learn the skills they would

The female cub
encounters her father –
a rare event in the lives
of tigers, as the male
has nothing to do with
raising his family.

need when they finally left her care. So after she had caught her victim, she would leisurely feed before they got their turn.

The differences between the four cubs were also becoming more and more apparent. There is normally a dominant cub in each litter – usually a male. But in this family, there were three males, plus a female named Biba.

For the few weeks after they were born they all looked more or less alike, but by the time they reached nine months old the males were considerably larger and stronger than their female sibling. That meant that she no longer got her full share of the food, and was always last in line for feeding. Only once they had eaten their fill, and retired for a siesta, could she finally feed unchallenged.

ONE DAY, BIBA was feeding on the remains of a kill while her brothers and mother slept. But as she did so, there was real danger nearby: a huge male tiger, patrolling right through the middle of Raj Bhera's territory, very close to the cubs. Not even Raj Bhera could fight off a determined male if he wanted to kill her offspring.

Tigers are – unlike most other members of the cat family – very good swimmers; indeed, they bathe regularly, especially to cool down on hot summer days. Raj Bhera and her cubs had their own favourite pool, and it was here that the male tiger had decided to take a dip.

With her mother and brothers asleep, digesting their meal, Biba finished feeding and also headed to the pool – not to bathe, but for a drink. As she approached, she caught an unfamiliar scent; and although she was not fully aware of the implications of this, she knew something was different from normal.

Curious, like most young tigers, she approached the male – an act that could have been very dangerous – perhaps even fatal. But fortunately, he recognised her as one of his own cubs. She was very lucky indeed: he was the only tiger in the whole of Bandhavgarh who would not kill or harm her. Male tigers rarely bump into their young, and when they do, they simply move on. It's a solitary life for a mature male tiger, as he constantly patrols his territory against rival males.

This was also a breakthrough moment for the crew: all that hard work setting and checking the camera traps had finally paid off, with a scene that would be almost impossible to have filmed without the remote cameras, as John Brown notes. "This was one of the most memorable moments of all – we had put out our camera traps wherever we could, and became increasingly tuned into the best locations. But the camera traps often failed to work for technical reasons, and checking the memory cards was always a heart-in-mouth moment. So to have this incredible encounter play out in front of the camera – and for the camera to work exactly as it should – was very special, and provided a unique insight into tiger behaviour."

Territory

BELOW

A tiger carrying one of its smaller prey items: a monkey.

Soon after all four cubs reached their first birthday – a real achievement for Raj Bhera – winter moved rapidly through the brief Indian spring and into the heat of summer. This brought widespread drought, which was good news for this tiger family, as the lack of rain meant that prey began to concentrate in large numbers around any remaining areas of standing water.

As well as the usual deer and wild boars, other animals not normally within the tigers' grasp – such as langur monkeys – now became part of their regular diet. These acrobatic primates usually spend most of the time out of the tigers' reach, high in the tree canopy, where they are safe from attack. But each summer, as the drought takes hold, the monkeys are desperate for water, and so they need to descend to the forest floor to drink. When they do so, they are naturally very cautious: the adults come down to the ground first, and only when the coast is clear do the youngsters follow. Yet however alert the monkeys are, a crafty tiger like Raj Bhera can still sometimes catch them. She usually did so by hiding in the long grass, watching and waiting for the moment a monkey would be within reach. Then she chased the group of monkeys until she caught a weak or young animal.

The time the cubs would be spending in their mother's care was rapidly running out. From roughly one year old, their permanent canines start to come

Raj Bhera's three male cubs, fast asleep after feeding.

OVERLEAF

The matriarch Raj Bhera with her four cubs: three males and the smaller female, Biba.

through, and this – combined with a rapid increase in their weight and size – enables them to start to hunt and kill smaller prey on their own. By now they had learned to stalk a range of animals including wild pigs, deer and birds; little by little, they were learning the skills they would need to hunt on their own.

Some time around the age of two the cubs would become independent and leave their mother, but most would stay in the area near where they were born for another year or so. Male cubs usually leave earlier than females, and will travel much farther: sometimes as far as 300 km (185 miles) away from where they were born. In contrast, females often stay close to their mother's territory – or even overlap it in part – for the rest of their lives.

In one study of dispersal in Chitwan National Park in Nepal, female cubs only moved on average 10 km (6 miles) away from their mother; while males went as far as 65 km (40 miles) away. All those females managed to establish their own territory, but fewer than half the males did so.

THE FIRST – and most important stage – in a tiger's adult life, once it has left its mother, is to establish its home range or territory. Here again, males and females pursue very different strategies. Although both lead mainly solitary lives, a female's territory is usually around 20 square km (eight square miles), whereas males' territories are three to five times larger, covering an area of 60–100 square km (20–40 square miles), to give them a better chance of finding one or more females with which they can breed.

'Scent-markings are a kind of coded message, imparting a whole range of information.'

However, these are not discrete territories: those of the males overlap with those of one or more females. And, as the team discovered during three years of watching and filming, here in Bandhavgarh, the females' territories were far more fluid than they had thought would be the case, shifting their boundaries and shapes from week to week. This is likely to be because the density of tigers here is closer to the limit they can achieve, so each female's territory is under greater pressure than in 'normal' situations.

Once a female tiger has chosen a territory she will normally stay in the same rough location for the rest of her life; males rarely get that choice, as they must constantly defend their territory against rival males, so the turnover of male territories is much higher.

A tiger's range is, like that of all territorial wild creatures, closely linked to the availability of its prey. So here in Bandhavgarh, where there are plenty of deer and wild boars, the ranges are on the smaller side. But in less favourable habitats, where prey animals are more dispersed, such as the forests of the Siberian Far East, they are much larger: the average Siberian (or Amur) tiger territory covers roughly 1,400 square km (540 square miles), many times larger than that of a male Bengal tiger.

And some tigers will venture much further. Siberian tigers may travel as much as 1,000 km (620 miles) in search of prey, simply because in such a cold climate, a tiger needs a huge territory in order to find enough to kill and eat. Bengal tigers rarely travel anything like as far; if they did they would soon run into issues with India's massive human population.

TIGERS, LIKE ALL big cats, have a very highly developed sense of smell, which they use when holding a territory. A male tiger will liberally spray his urine, and secretions from a gland at the base of his tail, at key points around his home area – usually over prominent rocks and trees; he will also make marks with his claws on the trees, and on the ground through the forest. These are clear warning signs to any intruding male, letting him know that this territory is already occupied, and encouraging him to move elsewhere.

Female tigers mark their territory too; this allows males to find them, especially when they come into heat, when their scent subtly changes, offering an invitation rather than a warning. All these scent-markings are a kind of coded message, imparting a whole range of information, such as the marker's sex and individual identity – for each scent is unique to that tiger.

Male tigers have different issues: especially when young. In an area such as Bandhavgarh, where tigers live at a very high concentration in a relatively small place, winning a territory in the first place is very hard indeed. In such circumstances, a young male must be patient, living on the edge of an older male's territory until either that male dies and there is a vacancy, or until he grows big and strong enough to oust the incumbent.

A sambar deer – one
of the most regular
prey animals of the
Bengal tiger.

Biba and a male cub
relax together in a shady
jungle glade.

But this is a risky business: not only might he be injured in any fight with a rival, but also he may not be able to find enough food, as the larger male wins the best kills. That explains why of all tiger populations, it is these transient young males that die at the highest rate – as many as one in three will not survive each year.

LIKE OTHER TOP predators, tigers are keystone species: without their presence, the whole ecology of their habitat would be changed irrevocably. This is not, as is often supposed, because they have an effect on the total numbers of their prey – they don't. Instead, it is because they create what scientists call the 'ecology of fear'.

Where top predators such as tigers are present, their prey is always on the move, for to stand still and feed in one place for too long would be to invite attack. This means that instead of areas of habitat being over-grazed or over-browsed, the vegetation always has the opportunity to renew itself. The healthy population of tigers such as Raj Bhera in Bandhavgarh is itself reflected in a healthy ecosystem, home to many other wild creatures.

Conversely, where predators such as tigers are removed from an ecosystem, and grazing animals no longer feel that they are constantly at risk of being attacked, they tend to stay in one place, and so cause more habitat destruction. So tigers are a key element in preserving the biodiversity of the forest ecosystem, to the benefit of all species.

Saving the tiger

OPPOSITE

Tourism is a way of bringing in money to help conserve tigers, but it has to be strictly controlled. The big cats seem to quickly become accustomed to the camera-toting visitors.

OVERLEAF

One of Raj Bhera's male cubs resting. During daylight hours, tigers spend much of the time resting or asleep.

In the past – indeed as recently as the late nineteenth century – tigers could be found across a wide expanse of the Eurasian landmass, from Eastern Europe in the west (around the Caspian and Black Seas) to the Amur Peninsula in the east; north to within reach of the Arctic Circle, and south beyond the Equator, to the islands of Sumatra and Java.

Today, their range has contracted massively, by over 90 per cent. They can no longer be found in Eastern Europe, Central or Western Asia, Bali, Java, and much of Southeast Asia. Even in their former stronghold of the Indian subcontinent, their range is far smaller than it used to be a century ago.

Whereas eight different races – or subspecies – are known to science, only five still survive; those distinctive animals from Bali, Java and Southwest Asia are now extinct, and three of the remaining five races (Sumatran, Amur, and South China tigers) are classified as Critically Endangered. The remaining two races – Bengal and Indo-Chinese – are slightly more numerous and widespread, but are nevertheless still classified as Endangered. The tigers in peninsular Malaysia may also be a separate race; if so, they too would be classified as Critically Endangered.

Numbers of tigers have been in freefall too. From an estimated global population of about 100,000 individuals in the year 1900, soon after the start of the new millennium that figure had dropped to just 3,200. In the early part of the twentieth century, the decline was mainly down to tiger-shooting. The ruling maharajahs and their hunting parties – often including British, American and European dignitaries, businessmen and even members of the British royal family – would bag dozens of tigers in a single day. Two maharajahs boasted that they had each killed more than a thousand tigers during their lifetimes.

Later, as tiger numbers declined, the hunting more or less stopped, and India's tigers were given stronger legal protection. This was at least partly thanks to the country's first female Prime Minister, Indira Gandhi, who was a great supporter of tiger conservation.

However, even though they were protected, illegal killing was rife, fuelled by the growing demand for tiger products in China. In the early 1990s, a survey in the famous Ranthambore National Park, known as a hotspot for tigers, found just 15 individuals – poachers had killed the rest, in order to supply the lucrative trade in tiger skins and other body parts. Soon afterwards, the authorities in Delhi seized a haul of almost 500 kg (1,100 lbs) of tiger bones. Poaching has been a major reason for the rapid fall in the tiger population, both in India and elsewhere in Asia.

'Today, tigers' range has contracted massively,
by over 90 per cent.'

Ironically, there are now far more tigers in captivity – at least 13,000, the vast majority in the United States – than there are in the wild. Today more than half of the world's wild tigers – at least 2,200 – are found in India, with smaller but still significant populations in the Russian Far East, Indonesia and Malaysia. Elsewhere in Asia, tigers are just about managing to hang on in Bangladesh, Nepal, Thailand, China, Bhutan, Myanmar, Laos and Vietnam. But most of these populations are small and highly fragmented, meaning that India's tigers provide easily the best hope for the continued survival of this magnificent species.

THE TIGER POPULATION may now have turned the corner and begun to slowly increase, but this should not obscure the fact that tigers are still in big trouble. Indeed, unless the current conservation plans succeed, tigers may conceivably go extinct in the wild within our lifetimes.

The reasons that tiger numbers have been in freefall for so long are many and complex. Perhaps the most serious issue is competition between tigers and humans for land: in India, whose population of more than 1.3 billion people is almost as many as China (whose land area is nearly three times the size), this is a perennial problem, especially as many tiger reserves are close to heavily populated areas. When tigers venture outside their protected areas, and kill livestock – or worse still, people – it is understandable that they are treated with fear and hostility.

The overall area occupied by all wild tigers, throughout their world range, has been estimated at less than 1.1 million square km (400,000 square miles), an area roughly the size of Ethiopia. This is only just over half the area tigers could be found as recently as the middle of the 1990s, and just seven per cent of the species' historical range.

More people inevitably lead to more habitat destruction. Even when this does not occur, a growing human population and its associated development of land often causes habitat fragmentation. Big predators such as tigers not only need large ranges, they need these areas to be continuous, uninterrupted habitat. These discrete areas then need to be connected to one another through corridors, down which the tigers and their prey can travel to suit different weather and varying feeding conditions through the seasons of the year. Habitat fragmentation doesn't just cause problems for tigers; it can also reduce the numbers and range of their prey.

Another huge and growing problem is the increasing demand for tiger skins, teeth, bones and other body parts, mainly to serve the demands of traditional medical practitioners in China or, in the case of tiger skins, for rich people to display as trophies. As a result of this demand, illegal poaching is widespread. During the first decade and a half of the twenty-first century, almost 1,600 tigers were taken from poachers and the illegal wildlife trade – equivalent to almost half the total world population.

'Unless the current conservation plans succeed, tigers may conceivably go extinct in the wild within our lifetimes.'

Wildlife tourism –
especially photography
– is a growing business
in India, and may help to
save the tiger.

In 2010, the umbrella conservation body the Global Tiger Forum was established. Tiger experts from all over the world came together to try to stop, or at least reduce, the current illegal trade in tiger products. They do so by using a carrot-and-stick approach: promoting anti-poaching measures on the ground, with the help of national and local governments, and working closely with local communities to get the people on the tigers' side; while at the same time trying to change cultural behaviour in China – by far the biggest market for tiger products.

And as if habitat loss and fragmentation, and illegal poaching, weren't enough, now climate change is becoming a major issue, especially for a predator that is so dependent on the continued healthy populations of its prey.

In India, a programme of habitat restoration and the improvement of nature reserves has begun to help restore tiger populations; and may explain the recent reported rise in numbers. Wildlife tourism – including the many thousands of visitors who come to see Raj Bhera and the rest of Bandhavgarh's tigers – is also a major asset, enabling the authorities to persuade local communities that, in the longer term, tigers are more important, and more economically valuable, alive than dead.

Conservation scientists have now identified five major areas of land in India, which collectively cover an area of almost 150,000 square km (roughly 58,000 square miles). Each of these areas is considered to be capable of providing enough food for 200 tigers – about a thousand individuals in all. Overall, the plan is to double tiger numbers in the wild by 2022: an ambitious target, but one certainly worth aiming for.

Solo returns

BELOW

A male sambar deer may weigh up to half a tonne – quite a challenge, even for a fully grown tiger.

For the tigers of Bandhavgarh, the dry season is a time of ease and plenty. Raj Bhera was fortunate in the timing of her cubs' birth; at the point when they had grown bigger, and needed more food, there was a glut of potential prey.

Sambar deer are always a favourite for tigers, because of their large size. With a shoulder height of 1.6 metres (over five feet), and weighing up to half a tonne, a male sambar is one of the world's largest deer – only the elk and moose are bigger.

By this point, each time Raj Bhera made a kill the three male cubs came in to feed straightaway. This left little or nothing for their smaller sister Biba, who only had a few scraps of meat to feed on, after her siblings had eaten their fill.

The time spent feeding was very much shorter than when the cubs were younger: "It was incredible how quickly the cubs' appetite increased", says John

ABOVE

Raj Bhera's daughter
Biba eventually left her
mother's territory, and
struck out on her own.

Brown. "A big sambar that would previously have lasted the family two or three days was soon being consumed in one sitting! That meant that the pressure on Raj Bhera was becoming relentless."

Meanwhile, if Biba were to have any chance of surviving into adulthood, and one day raising a family of her own, she had to learn to kill for herself – and fast. All tigers – like other big predators – practise hunting from an early age through their play: first by 'stalking' each other, or their mother, and then later, as they grow older, stronger and more experienced, by chasing any smaller animals such as rodents or small deer that they come across on their travels.

This stage in their life is always a critical time for female cubs, and many of them fail to survive into maturity, as they simply find it impossible to get enough to eat before they have developed the skills to hunt properly.

For Raj Bhera as well as Biba, things were soon to turn much tougher. The monsoon season was about to break, bringing constant and heavy rain for months on end, from the middle of June through to September. This glut of water allows the prey animals to desert the waterholes and forest ponds, and disperse far and wide across the whole of the park. Indeed, the rains are so heavy at this time of year that the park is closed to the public for several months.

With the whereabouts of her prey so unpredictable, Raj Bhera had to venture to the furthest bounds of her home range, so that she could find enough food to stave off starvation. It was on the western edge of her range that she picked up a familiar smell – and one that clearly meant trouble – that of her adult daughter, Solo.

During the time that Raj Bhera had been focused on keeping her cubs fed, staying in the heart of her range, Solo had taken advantage of her absence

> ' Raj Bhera was feeling the immense strain of providing for her cubs while at the same time trying to repel her daughter Solo. '

BELOW

Villagers watching out for Raj Bhera after she left the safety of the national park.

OVERLEAF

Raj Bhera, sitting in the meadow beneath the shadow of Bandhavgarh fort.

and moved into a large swathe of her territory. This not only put pressure on Raj Bhera, but it also meant that Biba, at seventeen months old, was being outcompeted by her larger and stronger male siblings for food. So she had no choice but to move away and find a territory of her own – much earlier than she would normally choose to do so. Sadly, Biba disappeared soon afterwards.

Raj Bhera was feeling the immense strain of providing for her cubs while at the same time trying to repel her daughter Solo. "When we first met her, Raj Bhera was in glorious condition," says Theo Webb. "But towards the end of the filming, she had become thinner and thinner, and her fur was looking rougher. You could tell she was having a really tough time."

Then Raj Bhera made the decision to leave the sanctuary of the park and head towards a local village – one of about 60 settlements supporting 40,000 people in the area around the park borders. Here, she inevitably caused panic and mayhem, as the villagers gathered around the tiger, beating the scrub with sticks and shouting at her, in a desperate effort to see her off. Their response was understandable: few wild creatures are quite as dangerous as a cornered tiger, and many people fall victim to them each year, especially in rural areas of India.

As soon as they heard that Raj Bhera was outside the park, Theo and the team raced to the village. Theo was very concerned with that they saw. "She had been pushed out of her territory by Solo, and in order to find food she'd

'By reclaiming her territory from her own daughter, Raj Bhera had re-established herself as the Queen of Bandhavgarh's tigers.'

come to a really risky area, to hunt cattle, dogs, or whatever she could find. This was a really tricky time for her."

As the park's tigers become more and more successful and numbers rise, causing more territorial rivalries between animals, some are inevitably going to be forced to venture outside the park in search of food. Conflict with humans presents a real danger to both sides, with potentially fatal results for both humans and tigers.

Fortunately, just when things were about to get very nasty indeed, the rangers from the national park arrived. "The tiger had been cornered in a patch of bamboo, where the park rangers had managed to tranquilise her, so they could take her back to her home territory – a real race against time, before the effects of the tranquiliser wore off. But this incident just showed the lengths Raj Bhera had to go to, so she could find enough food for herself and her cubs. This is the life for a modern tiger – and it certainly isn't easy."

Meanwhile, Raj Bhera's absence could have given Solo the perfect opportunity to move in and take over her mother's territory permanently. Fortunately, she did not have time to do so, as the park authorities soon released Raj Bhera back into her home area, where she could resume her support of her male cubs.

It was a heart-stopping moment. "Seeing Raj Bhera back in her home territory, after the frantic scenes outside the park, was a huge relief to us all," says Theo. "Inevitably when you spend so much time with these huge, charismatic animals you can't help becoming emotionally attached; so to see her life in jeopardy like that brought home to us the reality of the threats facing India's tigers."

By reclaiming her territory from her own daughter, Raj Bhera had re-established herself as the Queen of Bandhavgarh's tigers. She had successfully raised three out of her four cubs to adulthood, and close to being independent. Hopefully now she can continue her dynasty, and raise another generation of Bandhavgarh's tigers.

Wildlife cameraman John Brown, who along with director Theo and their team of spotters had spent more than 200 days in the field filming this magnificent beast, reflects on what the team learned about tigers through their time with Raj Bhera and her family. "I'd never spent such a prolonged period filming a predator before, and it was a wonderful experience to gain some sort of insight into Raj Bhera's character. She was skilful and incredibly tenacious, but she was also up against very tough odds. By the end of the three years, I could usually predict when she would stand up after a sleep, what route she would take into the meadow, what time in the evening she would start to move – but she would still keep us guessing – there are as many ways of being a tiger as there are tigers."

Index

Acknowledgements

Thanks to the whole *Dynasties* team, and to Steve Tribe, Beth Wright and Laura Barwick.
Stephen Moss

PRODUCTION TEAM

Sir David Attenborough

Tom McDonald

Adalean Coade
Alison Brown-Humes
Karen Hooper
Kirsty Emery
Lisa Sibbald
Louis Rummer-Downing
Loulla Wheeler
Luke Ward
Michael Gunton
Miles Barton
Nick Smith-Baker
Rebecca Hathway
Rosie Thomas
Rupert Barrington
Simon Blakeney
Theo Webb
Will Lawson

CAMERA AND SOUND TEAM

Alex Page
Barrie Britton
Bill Rudolph
Dave McKay
Ian Llewellyn
James Reed
John Aitchison
John Brown
Justine Evans
Lindsay McCrae
Luke Barnett
Mark MacEwen
Mark Yates
Mateo Willis
Matt Drake
Nick Lyon
Paul D Stewart
Rolf Steinmann
Sophie Darlington
Ted Giffords
Tim Shepherd
Warwick Sloss

POST PRODUCTION

Films at 59
Miles Hall
Wounded Buffalo

MUSIC

Benji Merrison
Will Slater
Dan Brown
Natasha Pullin

FILM EDITORS

Andy Mort
Angela Maddick
Dave Pearce
James Taggart
Matt Meech
Nigel Buck
Rob Davis
Robbie Garbutt
Robin Lewis

ONLINE EDITOR

Franz Ketterer

DUBBING EDITORS

Angela Groves
Kate Hopkins
Tim Owens
Ben Peace

DUBBING MIXERS

Chris Domaille
Graham Wild

COLOURIST

Simon Bland

GRAPHIC DESIGNER

Mick Connaire

BBC STUDIOS SALES AND DISTRIBUTION

Mark Reynolds
Patricia Fearnley
Amy Dowsett
Monica Hayes
Hayley Moore

SCIENTIFIC ADVISORS

Gerald Kooyman
Daniel Zitterbart
Jill Pruetz
Ullas Karanth
Esther van der Meer

FIELD ASSISTANTS, GUIDES & LOCAL EXPERTS

Alex Naert
Dave Breed
Desiree Murray
Digpal Karmawas
Dhruv Singh
Nick Murray
Dondo Kante
Henry Bandure
Jacques Tamba Keita
Michel Tama Sadiakhou
Samuel Munene
Simeon Josia
Stefan Christmann

WITH SPECIAL THANKS

Ann E. Bowles
Barbara Wienecke
Ben Simpson
Brian Heath
Bursa
Bushlife Conservancy
Ceri MacLure
Christine Wesche
Colin Jackson
Daniel Noll
Dominic Grammaticas
Donna Williams
Dr Shivani Bhalla
Eberhard Kohlberg
Emma Brennand
Etty Varley
Felix Riess
Fongoli Savannah Chimpanzee Project
Gert Uys
Gordon Leicester
Grant Bayliss
Hannes Laubach
Harrison Nampaso
Hauke Schulz
Jemal Guerrero
Jo Haley
Justin Heath and The Mara North Conservancy
Kalyan Varma
Karen Nicholls
Katrina Bradley
Khadidiatou Ba
Krithi Karanth
Lala
Madhya Pradesh Forest Department - Bandhavgarh Tiger Reserve
Malle Gueye
Mana Pools National Park & Staff
Martin Tweddell
Matthew Torrible
Maximillian Merl
Mboule Camara
Meemendra Kumar
Mike Saunders
Mussa Lekwale
Neelam Singh
Nick Turner
Nigel Adams
Patrick Beresford
Patrick Musiza Isaiga
Paul Thompson
Peter and Catherine Blinston, Painted Dog Conservation
Ronny Lebrenz
Sam Rogers
Sanju
Shona Harris
Sina Loschke
Sonu
Sven Kruger
Tessa Worgan
The Alfred Wegener Institute
The drivers and staff of Governors Camp, Masai Mara
Tim Heitland
Toby Sinclair
Tom Crowley
Ursula Schlager
Vijay Singh
Zimbabwe Parks and Wildlife Management Authority
Zsofia Juranyi

In memory of Tash Breed and Jean Hartley

Picture credits

Front cover BBC/BBC Studios/Stefan Christmann/Nick Lyon/Theo Webb
Back cover (l–r) Louis Rummer-Downing; Mark MacEwen; Theo Webb; BBC; Stefan Christmann

1 Mark MacEwen; **2-3** Louis Rummer-Downing;
4-5 Theo Webb; **7** BBC; **8-9** Nick Lyon

1 LIONS

All © Simon Blakeney except:
16 Federico Veronesi; **22-3** Federico Veronesi;
25 Louis Rummer-Downing; **28-31** BBC;
35 Louis Rummer-Downing; **40** BBC;
41 Mark MacEwen/naturepl.com;
42-45 BBC; **48** Denis-Huot/naturepl.com;
50 BBC; **52-3** Federico Veronesi;
55t and **m** BBC; **55b** Mark MacEwen;
60-1 BBC

2 CHIMPANZEES

72-3 Rosie Thomas; **74** Mark MacEwen;
75-7 John Brown; **78-9** Mark MacEwen;
80-1 Frans Lanting/FLPA; **82** Mark MacEwen;
83 John Brown; **84** Rosie Thomas; **86** John Brown;
87t BBC; **87b** Mark MacEwen; **88** John Brown;
89 Mark MacEwen; **90** John Brown;
91 Fiona Rogers/shahrogersphotography.com;
92 Mark MacEwen;
94 Fiona Rogers/shahrogersphotography.com;
95 John Brown; **96** Rosie Thomas;
97 Mark MacEwen;
98-9 Fiona Rogers/shahrogersphotography.com;
100 John Brown; **102** BBC; **103** John Brown;
105-6 Mark MacEwen; **108-9** BBC;
110 Rosie Thomas; **111-4** Mark MacEwen; **115** BBC;
117 Mark MacEwen; **118** Suzi Eszterhas/Minden/FLPA;
119 BBC; **121** Mark MacEwen; **122-3** John Brown

3 PAINTED WOLVES

All © Nick Lyon except:
140 BBC; **144** BBC; **150-1** BBC;
157 BBC; **162-4** BBC

4 PENGUINS

All © Stefan Christmann except:
185 Lindsay McCrae; **186-7** Will Lawson;
214-5 Paul Nicklen/Getty; **222-3** BBC;
227 BBC;

5 TIGERS

All © Theo Webb except:
235 John Brown; **241** BBC;
246 Vivek Sharma/Minden/FLPA; **249** BBC;
252 BBC; **253** Louis Rummer-Downing;
261b BBC; **262** Andrew Parkinson/naturepl.com;
271 Andy Rouse/naturepl.com;
275t Will Watson/naturepl.com; **275b** John Brown;
276 Andy Rouse/naturepl.com

endpaper *front* Stefan Christmann
endpaper *back* Nick Lyon

3 5 7 9 10 8 6 4 2

BBC Books, an imprint of Ebury Publishing
20 Vauxhall Bridge Road,
London SW1V 2SA

BBC Books is part of the Penguin Random House group of companies whose addresses
can be found at global.penguinrandomhouse.com

This book is published to accompany the television series entitled *Dynasties*
first broadcast on BBC One in 2018.

Executive producer: **Mike Gunton**
Series producer: **Rupert Barrington**

First published by BBC Books in 2018

www.penguin.co.uk

A CIP catalogue record for this book is available from the British Library

978-1-785-94301-0

Commissioning Editor: **Albert DePetrillo**
Project Editor: **Bethany Wright**
Picture Research: **Laura Barwick**
Image grading: **Stephen Johnson, www.copyrightimage.co.uk**
Design: **Bobby Birchall, Bobby&Co**

Printed and bound in Italy by Printer Trento

Penguin Random House is committed to a sustainable future for our business, our readers and our planet.
This book is made from Forest Stewardship Council® certified paper.